Integrated Use of Urea-DAP Briquettes for Crops

Krishiratna Dr. Narayan K. Savant

Retd. Soil Scientist
International Fertilizer Development Centre
Alabama (USA)

Dr. Kashinath R. Sonar

Retd. Head
Deptt of Soil Science
Mahatma Phule Agricultural University
Rahuri (Maharashtra) India

ibdc publishers
PUBLISHING GREEN

Published by

ibdc publishers
PUBLISHING GREEN

Khushnuma Complex Basement
7, Meerabai Marg (Behind Jawahar Bhawan),
Lucknow 226 001 U.P. (INDIA)
Tel. : 91-522-2209542, 2209543, 2209544, 2209545
Fax : 0522-4045308
E-Mail : ibdco@airtelmail.in

First Edition 2010

ISBN : 978-81-8189-560-8 (H.B.)

Composed & Designed at :

Panacea Computers
2nd Floor, Agarwal Sabha Bhawan, Subhash Mohal
Sadar Cantt., Lucknow-226 002
Phone : 0522-2483312, 9335927082, 9452295008
E-mail : prasgupt@rediffmail.com

Printed at:

Salasar Imaging Systems
C-7/5, Lawrence Road Industrial Area
Delhi - 110 035
Tel. : 011-27185653, 9810064311

Dr. Vijay Mehta
M.Sc. (Agri.), Ph.D.
VICE CHANCELLOR

DR. BALASAHEB SAWANT
KONKAN KRISHI VIDYAPEETH
Dapoli, Dist. Ratnagiri
(Maharashtra) Pin-415 712

Foreword

A great deal is being talked about on the agricultural advancement of the country underlining the increase in produce and productivity. However, that does not speak about with all the crops, as the productivity of most of the crops in India is comparatively low due to various reasons and one of the most important reasons is low fertilizer use efficiency. The fertilizer use efficiency in rice and other crops is about 20-40% for nitrogen. Thus, it causes a great loss, particularly to small and resource poor farmers. In this situation, this book "Integrated Use of Urea-DAP Briquettes for Crops" written by Dr. N. K. Savant and Dr. K. R. Sonar is published in good times.

In India, agriculture is experiencing a slow down on several facets, the area under various crops is reducing and food grain production is declining. With holistic approach to agriculture and agro-economy, Dr. Savant developed *Charsutri* (four points) Rice Agrotechnology, also known as Savant's Integrated Rice Agrotechnology for lowland and upland rice, wheat, sorghum and pearl millet crops. This technology is simple, easy to adopt and has many advantages. It will promote large-scale manufacture and wide use of Urea-DAP briquettes in future. The authors have verified this technology by conducting more than 3500 demonstrations / trials on farmers' fields in Maharashtra and Gujarat states through Janaseva Foundation (Silvassa), Pune. Dr. Savant has also developed few agricultural implements such as Paired-Row Seed Drill, Kranti Briquetter, Briquette Applicators useful for this technology. The technology could directly be used for food grain crops all over India and will certainly help making the farming profitable. The authors have also suggested use of Urea-DAP briquettes with drip irrigation system. They have used large number of illustrations to make their calling easy to understand. I would like to congratulate

Dr. Savant and Dr. Sonar for bringing out this informative and supportive book.

I am sure that, this book will be useful to all those related to agriculture - students, teachers, researchers, extension workers from agriculture departments, KVKs, NGOs, policy makers, fertilizer dealers and especially, the farmers.

May 11, 2009

Vijay Mehta
(Dr. Vijay Mehta)

Preface

Declining soil productivity of intensively cultivated areas of many developing countries is a serious problem threatening the food security of the regions. The tropical soils grown with food grain crops are generally low in organic matter, available N and P leading to widespread response to application of fertilizers N and P. Fertilizer N is extensively used by farmers in the form of urea, but its efficiency is very low due to improper method of application. Poor efficiency of fertilizer N by many crops is due to losses (up to 60% to 80%) from soil-plant system through ammonia volatilization, denitrification, run off and leaching.

Integrated use of different sources of nutrients (chemical fertilizers along with organic manures, crop residues, green manuring, biofertilizers etc.) has resulted in increasing the crop yields, maintaining the soil health and productivity. Organic-inorganic farming system uses chemical fertilizers in balanced proportion and limited quantity efficiently. Chemical fertilizers are not poisonous, but essential and useful for the crops.

Considering these points scientifically, the senior author has developed Urea-DAP briquettes and its agrotechnology for the betterment of the crops. These briquettes are simple to prepare, inexpensive, avoids pollution of soil, water and air and also useful for increasing crop yields. This integrated (*Charsutri* i.e. four practices or steps) organic-inorganic farming agrotechnology has been developed for transplanted and upland rice, wheat, sorghum and pearl millet crops. This technology, if used by Agriculture Department, Agricultural Universities, Krishi Vigyan Kendras, Non Government Organisations, Agripreneurs and Farmers, it will increase the production of the food grains. This will certainly minimise our food problem in future.

Agro One is a daily newspaper in Marathi published from Pune and completely devoted to Agriculture. It is widely circulated in Maharashtra and is very popular among the farmers and those related with agriculture. The senior author has published a book in Marathi "Urea-DAP briquettes for crops" in the year 2007 for the use of small farmers. So now we would like to make the same information available in English for the farmers and those related with agriculture all over India.

This publication is based on the research work carried out by the senior author for more than 30 years on various aspects of N and P availability in soil mainly for rice in India (Mahatma Phule Krishi Vidyapeeth, Rahuri), International Rice Research Institute (IRRI) in Philippines, and International Fertilizer Development Centre (IFDC) in USA. Later on the work was extended to other cereals viz., wheat, sorghum and pearl millet. Based on this research, the senior author has developed an Integrated (*Charsutri* i.e. four steps) Agrotechnologies for rice (transplanted and upland), wheat, sorghum and pearl millet crops. These agrotechnologies have been fully tested on the farmers' fields by conducting more than 3500 demonstration trials in Maharashtra and Gujarat States by Janaseva Foundation (Silvassa), Pune (India); State Department of Agriculture (Pune, Kolhapur, Nasik and Thane Regions); Zilla Parishad (Kolhapur and Ratnagiri districts); Dr. Balasaheb Sawant Konkan Krishi Vidyapeeth, Dapoli (Department of Soil Science and Agricultural Chemistry, Dapoli; Regional Agricultural Research Station, Karjat; Agricultural Research Station, Phondaghat and Khar Land Research Station, Panvel); Mahatma Phule Krishi Vidyapeeth, Rahuri (Agricultural Research Station, Igatpuri; Agricultural Research Station, Lonavala; College of Agriculture, Pune and College of Agriculture, Dhule); Central Rice Research Institute, Cuttack; International Rice Research Institute, Philippines; International Fertilizer Development Centre, USA etc. All the figures in this publications have been drawn by Dr. N.K. Savant.

The authors wish to extend their grateful acknowledgements to all the people who have conducted field demonstration trials, including the farmers. Special thanks are extended to Dr. S. Y. Daftardar, President, Janaseva Foundation (Silvassa), Pune and his colleagues for a very difficult task of conducting these numerous demonstration trials on the farmers' fields. The authors wish to thank Dr. Vijay Mehta, Vice Chancellor, Dr. Balasaheb Sawant Konkan Krishi Vidyapeeth, Dapoli, Dist. Ratnagiri (Maharashtra) for writing the foreword. Sincere thanks are due to Dr. S. B. Varade, Former Joint Director, WALMI Aurangabad; Dr. B.B. Desai, Former Professor and Head, Department of Biochemistry; Dr. S.Y. Daftardar and Dr. R.B. Somawanshi, Former Heads, Department of Soil Science and Agricultural Chemistry, Mahatma Phule Krishi Vidyapeeth, Rahuri for going through the manuscript and making useful suggestions. Thanks are also due to Prof. A.Vivian Thomas, Pune and Mr. Ashish A. Redkar, Ahmednagar for editing the manuscript and to Mrs. Asha Wadgaonkar for the excellent and neat typing.

We hope that this publication will be useful to Researchers, Scientists, Teachers, Agricultural Education Development Professionals, Students, Agriculture Department, Extension Workers, Krishi Vigyan Kendras, Agripreneurs, Non Government Organisations, Farmers and all those related to agriculture.

April 27, 2009 | N.K. Savant
Akshay Tritiya | K.R. Sonar

Contents

Chapter **1**

Brief History of Deep Placement of Fertilizers

Chemical fertilizers are poisonous for the crops and due to their use the soils become unproductive. Such statement is made by many organic farming promoters / followers. They insist not to use chemical fertilizers and promote that organic farming is the only alternative for sustainable crop production. Two main reasons for misunderstanding about the use of chemical fertilizers :

i) Fertilizer Industry personnel and Agricultural Scientists have started selling fertilizers without developing technology for the use of chemical fertilizers in the field.

ii) Most Indian farmers use chemical fertilizers and water indiscriminately and inefficiently (Photos 1 and 2). This can be improved by giving proper training to the farmers.

Photo 1

Photo 2

Photo 1 and 2. Inefficient use of chemical fertilizers for rice (Photo 1) and wheat crops (Photo 2)

Agricultural Scientists and the Fertilizer Industry have not developed technologies for the use of chemical fertilizers in the fields but the farmers are held responsible for the ill-effects of using fertilizers. Considering the situation in India, there is a need for researchers to develop agrotechnologies for the use of chemical fertilizers by the farmers. Urea is the main commercial form of nitrogen fertilizer used by Asian farmers (Stangel and Harris, 1987). Today, the rice farmers' practices of urea fertilizer application generally includes basal / broadcasting with or without incorporation of organic manure before transplanting and / or one or two topdressings under submerged condition immediately after transplanting up to the flowering (reproductive) stage. These conventional practices are largely based on common sense or convenience and are generally controlled by agroclimatic and socio-economic factors. Numerous research reports have demonstrated that these management practices for the application of fertilizer nitrogen in transplanted rice are very inefficient, and often results in 60-80 percent loss of applied nitrogen. These losses are mainly in gaseous form (ammonia volatilization and / or denitrification) from irrigated rice fields and run off and / or leaching from rainfed rice fields (Savant and De Datta, 1980 ; Vlek and Byrnes, 1986 ; Savant and Stangel, 1990). Nitrogen losses of this magnitude and nature are serious because they represent major financial losses to the farmers as well as nation's economy (Fig.1).

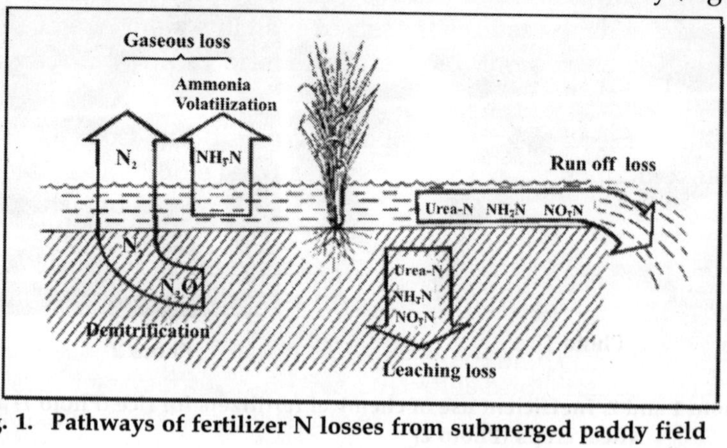

Fig. 1. Pathways of fertilizer N losses from submerged paddy field

Deep-point placement of nitrogen as urea supergranules (USG) or pillow (oval) shaped Urea-DAP briquettes after transplanting is an efficient nitrogen management practice in rice. But the conventional basal application of water soluble phosphorus (P), such as single superphosphate, DAP, is not an environment friendly fertilizer management practice, because P in the floodwater adsorbed by clays in the surface soil is prone to surface run off losses (Savant *et al.*, 1993 ; Daftardar *et al.*, 1997 b). Deep-point placement of chemical fertilizers in the root zone of soil is one of the efficient methods for the use of chemical fertlizers. A brief history and methods of the fertilizer use are given below.

1.1 Nitrogenous Fertilizers

Research on deep placement of fertilizer nitrogen for transplanted rice has been reported by Japanese Agricultural Scientists during 1930s. Since then agricultural scientists have begun to work on the use of ammonium sulphate, liquid ammonia and then urea as a straight fertilizer or supergranules for transplanted rice. These experiments were of two types.

i) Deep-non point placement of nitrogen fertilizer powder in dry soil before transplanting of rice or in wet soil after transplanting of rice.

ii) Deep-point placement of nitrogen fertilizers in the form of small or big pellets at transplanting or after transplanting of rice.

Irregular and untimely supply of canal water and wrong placement of nitrogen fertilizers (powder) are the two important reasons for not becoming the above two methods of fertilizer application popular among Japanesefarmers, even in improved rice cultivation. After that scientists carried out several experiments for deep placement of nitrogen fertilizer.

Deep placement of nitrogen fertilizer for rice was practiced in Japan during the early 1930s and in India during the 1950s. Since then, researchers from a number of national and international institutions have led to development and evaluation of concepts for rice cultivation in various parts of the world. Out of these information on few selected developments of the current agrotechnologies of deep placement of urea supergranules (USG) and Urea-DAP briquettes in transplanted rice are given below.

1930 : Japanese farmer, named Kenechi Sato deep-placed N fertilizer in the rice fields for the first time (Watanabe and Mitsui, 1979).

1942-52 : Japanese scientists reported enhanced efficiency of deep-placed ammonium sulphate, resulting in increased rice yields by 10% (Mitsui, 1977).

1952 : Indian scientists used pellets of N fertilizer + mud mixture in transplanted rice fields (Vachhani, 1952).

1955-58 : Japanese scientists reported increased efficiency by deep-placing prilled urea fertilizer in dry soil by ploughing or harrowing before irrigation (Watanabe and Mitsui, 1979).

1961-62 : Japanese scientists developed hand-operated implement / applicator for deep placement of urea (^{15}N) in soil and established minimum losses of nitrogen (Kunitaka, 1962).

1962-68 : Nitrogen studies showed that deep placement of prilled urea increased uptake of fertilizer nitrogen by rice plants (De Datta *et al.*, 1968 ; IAEA, 1970).

1973 : Chinese scientists developed a machine for preparing one gram pellets of ammonium bicarbonate for the supply of N in rice fields (Li and Chen, 1980).

1974 : International Rice Research Institute (IRRI) in Philippines worked on ^{15}N + mud pellets using mudball technique for the supply of nitrogen to transplanted rice (De Datta *et al.*, 1974).

1975 : International Fertilizer Development Centre (IFDC) in USA developed urea supergranules (USG) for use in transplanted rice (Stangel, 1989).

1975 : Norsk Hydro Organisation in Norway developed urea pellets using pan granulation technique (Holte *et al.*, 1982).

1975 : IRRI in Philippines developed an applicator for deep placement of urea pellets in rice fields at the time of ploughing (IRRI, 1977).

1977 : Indian Farmers Fertilizer Co-operative Ltd. (IFFCO) prepared one gram urea supergranules (USG) on a large scale for rice crop and conducted large number of field trials on farmers' fields all over India (Kumar *et al.*, 1989).

1978-79 : Dr. Savant worked at International Rice Research Institute (IRRI) in Philippines on urea supergranules (USG) in rice fields and reported results on availability of NH_4^+-N, diffusion of N from the point of placement and duration of N availability to transplanted rice crop (Figs. 2 and 3) (Savant and De Datta, 1980; Savant *et al.*, 1982).

1979 : Tamil Nadu Agricultural University (TNAU), Coimbatore, developed an applicator for deep placement (5-7cm) of pellets containing fertilizer + insecticide mixtures in rice fields (Ponnaiyya and Varadarajan, 1979).

1983 : Food and Agriculture Organisation (FAO) in Italy declared that deep placement of urea supergranules (USG) was useful for increasing rice yields (FAO, 1984).

1984 : Three organisations in Netherlands (Agriculture Bureau of the Netherlands Fertilizer Industry, the Department of Agricultural Engineering of Wageningen Agricultural University and the Dutch Farm Machinery Manufacturer Homeco B.V.) developed a pneumatic injector for deep placement of prilled urea in rice fields (Van der Sar and de Vries, 1984).

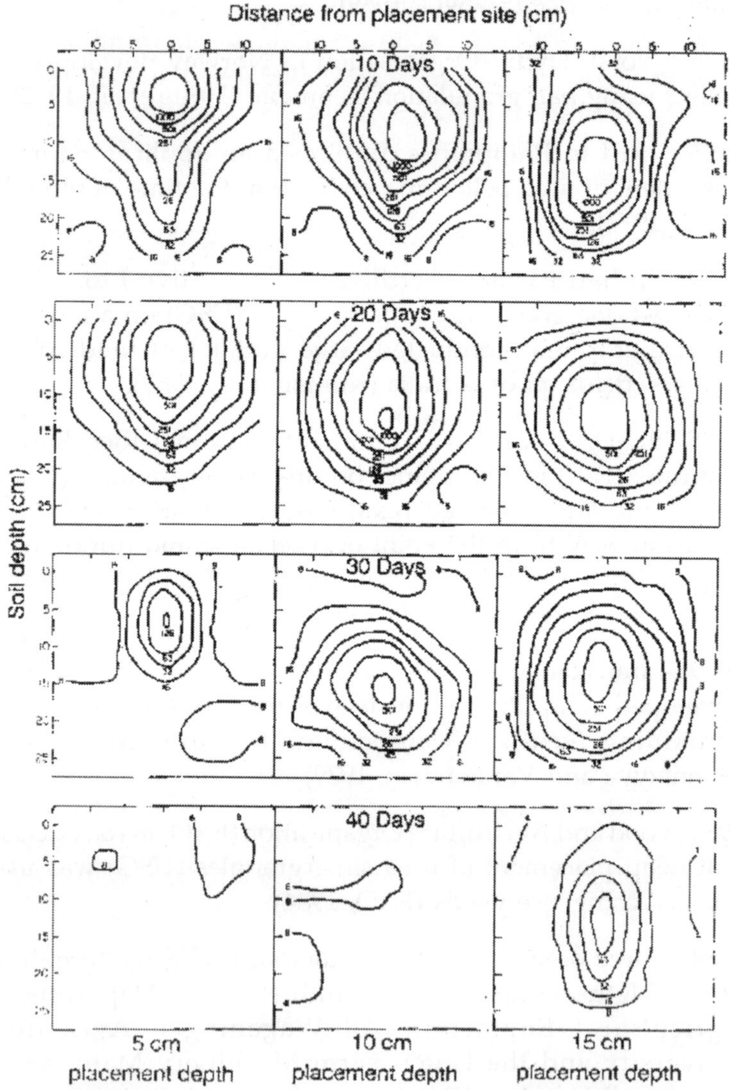

Fig. 2. Computed contour plots showing distribution patterns of ammonium-N (µg cm^{-3} wet soil) after placement of 2g USG at 10 cm depth in transplanted rice with square planting (20 cm x 20 cm) (IRRI, 1979 wet season)

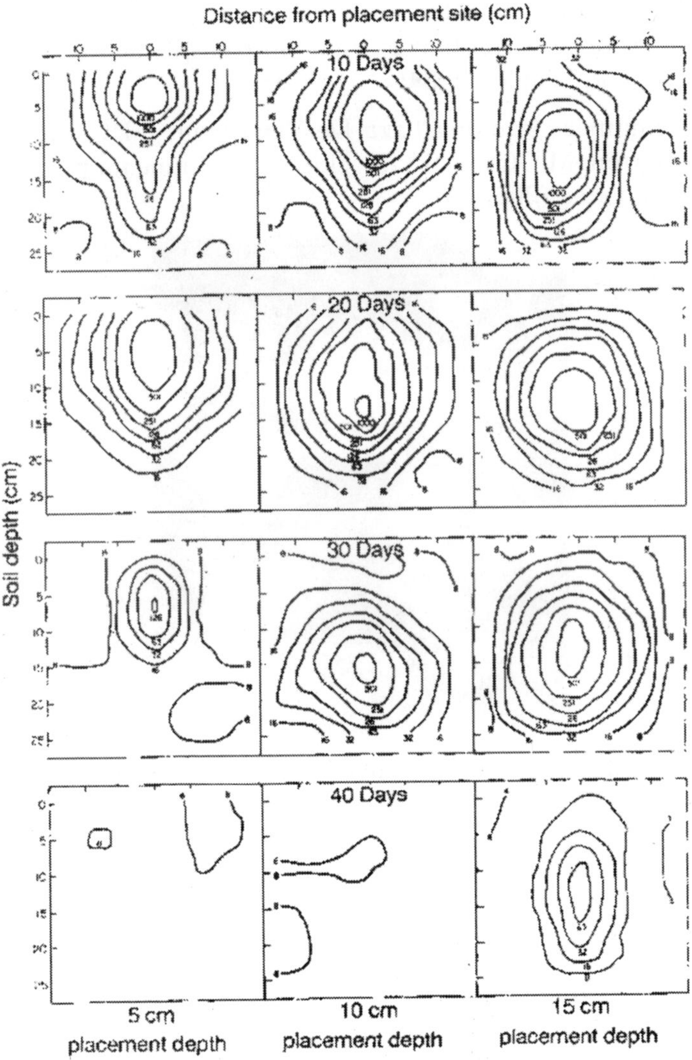

Fig. 3. Computer contour plots showing distribution patterns of NH_4^+-N µg cm-3 wet soil) after deep placement of 2g USG in a wetland rice having square (20 cm x 20 cm) planting geometry (IRRI, 1979 dry season)

1986 : International Rice Research Institute (IRRI) in Philippines developed liquid injector with a peristaltic pump for deep placement of urea solution in a line in transplanted rice fields (Bautista and Schnier, 1989).

1987 : Engineer Ramona Laza De Vega from IFDC and Inco Sumatra from MIDC in USA developed a machine for preparing urea briquettes (Photo 3).

Photo 3. Ramona Laza De Vega and Inco Sumatra developing briquetter machine (1987)

1990 : Savant and Stangel published a paper on "Deep placement of urea supergranules in transplanted rice : principles and practices". *Fertilizer Research*, 25 : 1-83, 1990.

1.2 Urea – Diammonium Phosphate (DAP) Briquettes

1990 : Savant and Chien from IFDC in USA published a paper on "Greenhouse evaluation of urea supergranules containing diammonium phosphate for transplanted rice". *Intern. Rice Res. Newsletter*, 15 (3) : 23 (1990).

1990-93 : Dr. Savant conducted a series of demonstration trials with the help of scientists from Dr. Balasaheb Sawant Konkan Krishi Vidyapeeth, Dapoli and Mahatma Phule Krishi

Vidyapeeth, Rahuri in Maharashtra (India) on their research farms and on the small farmers' fields using Urea-DAP briquettes technology for transplanted rice.

1993-95 : Janaseva Foundation (Silvassa), Pune conducted trials on transplanted rice using Urea-DAP briquettes scientifically on farmers' fields in tribal areas of Maharashtra.

1995 : Dr. Savant from Pune developed Savant's Integrated Rice Agrotechnology (SIRA) based on Integrated Organic-Inorganic Farming and use of deep placement of Urea-DAP briquettes (Practice 4 in *Charsutri*) (Savant, 2000 ; 2001a).

1996-99 : Scientists from Dr. Balasaheb Sawant Konkan Krishi Vidyapeeth, Dapoli, conducted demonstration trials on Urea-DAP briquettes for transplanted rice for three years. Based on these results Joint Agricultural Research Council (Agresco) of the four Agricultural Universities in Maharashtra recommended to Maharashtra State, use of Practice 4 (4th step) from SIRA (*Charsutri*) agrotechnology i.e. deep placement of Urea-DAP briquettes at the centre of square of four hills of rice in the field.

2000 : Department of Agriculture, Maharashtra State advocated use of SIRA for transplanted rice farming and decided to promote subsidy to buyers of "Kranti Briquetter" machine for preparing Urea-DAP briquettes.

2001 : Dr. Savant from Pune declared Urea-DAP briquettes as *Vikram*-2 briquettes at farmers' rally held at Kololi village (Tah. Panhala) in Kolhapur district.

2002 : A report of state-level seminar on Savant's Integrated Rice Agrotechnology was presented by Dr. Savant, in the seminar at College of Agriculture, Pune, March 7 (Savant, 2002).

2005 : Dr. Savant from Pune (2005) published a paper on "Savant's Integrated Rice Agrotechnology for Farmers of Maharashtra - A Review" in the Journal of Maharashtra Agricultural Universities (Savant, 2005).

1996 – 2005 : Janaseva Foundation, Pune conducted more than 3500 demonstration trials successfully on transplanted rice on farmers' fields in tribal areas of Maharashtra and Gujarat using Urea-DAP (*Vikram*-2) briquettes on a large scale.

2006 : Maharashtra Government honoured Dr. Narayan K. Savant, Pune by offering "Dr. Panjabrao Deshmukh Krishiratna Award" in 2005 which includes a gold medal, citation and a cash prize of Rs. 50,000/- for developing SIRA using Urea-DAP briquettes for transplanted rice efficiently.

2005–2007 : Janaseva Foundation (Silvassa), Pune has successfully conducted more than 3500 trials on transplanted rice and recorded over 40 q ha^{-1} rice production in tribal districts of Maharashtra and Gujarat States.

Chapter **2**

Manufacture of Urea-DAP Briquettes

Basically USG fertilizer is a simple physical modification of ordinary urea. It consists of large discrete particles of urea having 46% N as NH_2 (amide form). Weight may vary from 1 to 2 g per granule. Their shape depends on the production process. The USG prepared by any one of the melt type granulation processes are nearly spherical and without edges, but they may vary substantially in size and weight. Urea-DAP briquettes prepared by briquetting machine are pillow (oval) shaped with broken edges due to webbing and are reasonably uniform in size and weight. Biuret content is not a problem in USG prepared by briquetting or even by granulation because of adequate control on process variables such as temperature and retention time in granulation (Savant and Stangel, 1990).

Briquetting is a special type of compaction process that can be used for the production of urea briquettes or Urea-DAP briquettes. The basic principle of the compaction process is the application of pressure on Urea-DAP (prilled or fines) fertilizer as it passes between the two pocketed rollers rotating at the same speed but in opposite directions (Fig.4). The briquetting technology is simple and cost effective when compared with the melt granulation process (IFDC, 1989; Lupin, 1984).

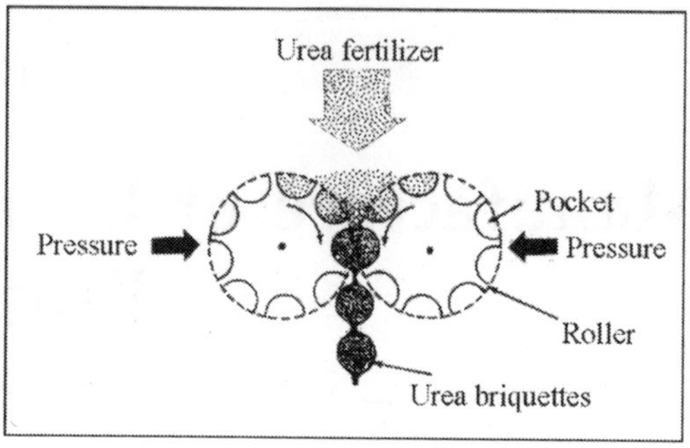

Fig. 4. Schematic diagram of roller in briquetter machine for preparing USG / Urea-DAP briquettes

The Fusian Academy of Agricultural Sciences and Yongitai Farm Machinery Factory from Peoples Republic of China developed a briquetting machine, for use in rural areas. With this briquetting machine, they manufactured ammonium bicarbonate briquettes and it is now used for manufacture of urea briquettes (Photo 4).

Photo 4. Chinese briquetter machine for use in rural areas

The IFDC in USA made some changes in Chinese Briquetter machine with the help of Metal Industries Development Centre and Centre for Soil Research from Indonesia and developed an improved briquetter (Photo 5). They evaluated the briquetter machine and technical drawings were made available to all the countries at US $ 700/-.

Photo 5. Dr. Savant, Dr. Chien and Mr. Ramona Laza inspecting 1 g urea briquettes prepared at IFDC, USA (1993)

Dr. N. K. Savant developed Savant's Integrated Rice Agrotechnology in 1993, in Maharashtra. At that time, Janaseva Foundation, Pune recommended Suyash Trust to purchase technical drawings of briquetter machine considering the demand for Urea-DAP briquettes from the farmers of Konkan region. Accordingly, Suyash Trust purchased technical drawings and developed Indian make Kranti Briquetter in consultation with Dr. Savant. After this, Messers Shanta Engineering, Kopari, (Thane district) started manufacturing Kranti Briquetters (Photo 6).

Photo 6. Preparation of Urea-DAP briquettes (about 2.7 g) using Indian make Kranti Briquetter (1995)

Kranti Briquetter is an indigenous equipment redesigned from the original model of IFDC, USA (IFDC, 1989; 1990). Kranti Briquetter is more efficient because it is designed to produce about 2.7g Urea-DAP briquettes, whereas the original IFDC model was designed to produce 1g briquettes. The briquetter is simple to use, easy to operate, portable, produces 200-250 briquettes per hour and operates on 5 H P electric motor (Savant, 2001a). With this Kranti Briquetter, Urea-DAP briquettes (*Vikram*-2) are obtained from 60% urea and 40% DAP having on an average 2.7 gram weight (27 g per 10 briquettes) (Fig. 5 and Photo 7). At present, Agriculture Department of Maharashtra Government provides subsidy on purchase of Kranti Briquetter.

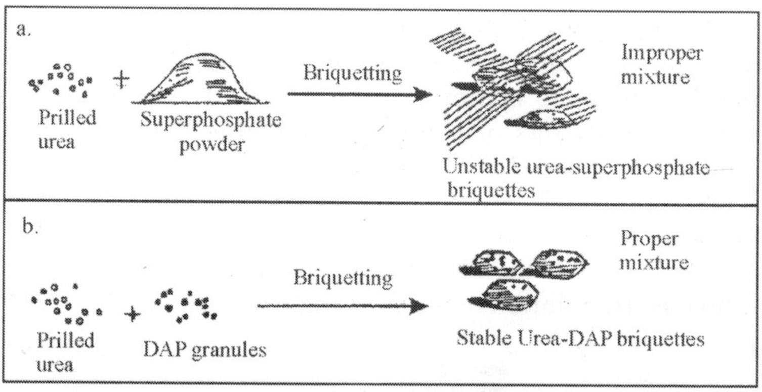

Fig. 5. Information about Urea-DAP briquettes

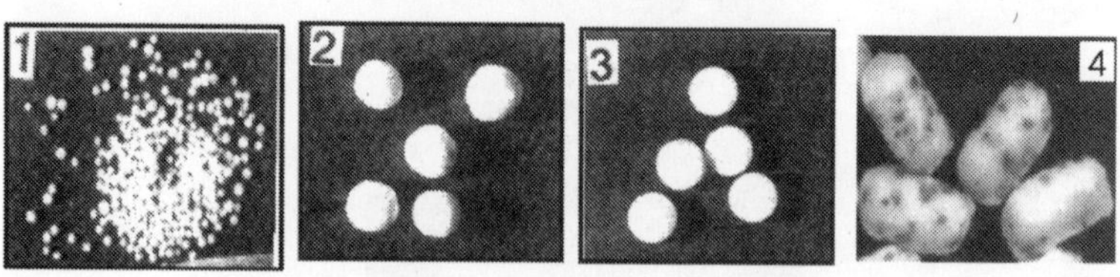

Prilled urea **1 g Urea supergranules (USG)** **1 g Urea pellets** **2.7 g Urea-DAP briquettes**

Photo 7. Different types of uera fertilizer

2.2 Important care about Urea-DAP briquettes

i. Urea and DAP fertilizers can be mixed together and used for preparing briquettes. But urea and superphosphate fertilizers cannot be blended together as it results in a new compound due to chemical reaction. Such a compound is unstable and cannot be used for preparing stable briquettes (Savant, 2001a). In short, urea and superphosphate fertilizer mixture is not suitable for preparing briquettes (Fig. 5).

ii. Urea-DAP fertilizer mixture being hygroscopic absorbs moisture from the atmosphere. Therefore, this fertilizer mixture should be used immediately for manufacture of briquettes. The briquettes should be stored in sealed good plastic bags up to 25 kg weight. These briquettes can be stored up to one year. Beyond one year, the briquettes may absorb moisture from the atmosphere (Savant, 2001b).

Chapter 3

Urea-DAP Briquettes for Crops : A Boon to Small Farmers

3.1 21st century farming needs

In times to come, Urea-DAP briquettes can be a potential source of N and P for rice and a boon to small farmers, particularly for the economically disadvantaged small rice farmers. Urea-DAP fertilizer briquette is a simple inexpensive physical modification of conventional urea and DAP and its improved management is technically a very sound practice.

Today, the improved management of Urea-DAP briquettes has the potential to meet the need of majority of small rice farmers in rainfed lowland and irrigated ecoregions where labour is available but cash resources are limited (Savant and Stangel, 1998). At present, propaganda for use of organic farming is being made by people who are agriculturists and / or politically influential. In fact in India, there is a need to use Integrated Nutrient Management through Organic-Inorganic Farming by farmers for increasing food grains production. This is because every year the area under agriculture is decreasing while farmers have to maintain the soil health and its productivity for increasing per hectare crop production to feed the ever increasing human population. In Maharashtra and other parts of the country, majority of the farmers are small and resource-poor having limited money for farming. If these farmers follow Integrated Organic-Inorganic Farming using Urea-DAP briquettes, the ill effects of conventional use of only chemical fertilizers could be avoided. At the same time, cost of cultivation (seeds, fertilizers and labour) could be reduced and still produce on an average 50% higher grain and straw yields to achieve profitable farming. This will certainly reduce the instability in food grains and make food security sustainable.

3.2 Brief Information about Briquettes

- **Name of Urea-DAP Briquettes**

Dr. N. K. Savant, Pune developed Savant's Integrated Rice Agrotechnology (SIRA) and Urea-DAP briquettes in the year 2001 (Savant, 2001a). He named Urea-DAP briquettes as *Vikram*-2 briquettes at the farmers' rally in Kololi village, Tah. Panhala, Dist. Kolhapur (Maharashtra).

- **Chemical Fertilizers**

Urea-DAP briquettes are prepared from the mixture of urea and DAP fertilizers in the proportion of60% urea and 40% DAP without using any binder.

- **Proportion**

Urea-DAP briquettes have 34% nitrogen (N) + 17% phosphate (P_2O_5), [2 nitrogen (N) + 1 phosphate (P_2O_5) , or 4 nitrogen (N) + 1 phosphorus (P)] (Savant and Stangel, 1995; 1996; 1998).

- **Weight**

Average weight of one Urea-DAP briquette is 2.7 g (27 g per 10 briquettes).

- **Shape**

The Urea-DAP briquettes are pillow (oval) shape (Photo 8).

- **Strength**

It is important that the briquettes should be stable and intact during handling and until they are placed in a paired-row in the field.

- **Agrotechnology for the crops**

Agrotechnology for integrated (*Charsutri* i.e. four practices/ steps) transplanted rice, drilled paddy, wheat, sorghum, and pearl millet cultivation is now available (Savant, 2007). But there is a need for development of agrotechnologies for use of briquettes for fruits, vegetables such as tomato, brinjal, chillies etc. and sugarcane for the farmers in the near future for increasing the yields.

- **Water management**

Instead of using expensive liquid fertilizers through drip irrigation, there is a need to develop a technology for integrated use of drip irrigation using Urea-DAP briquettes (see Chapter 13) for crops except submerged paddy.

- **Approval of Maharashtra Government**

Under the Fertilizer Control Order 1985, Maharashtra Government has notified an approval for use of Urea-DAP briquette fertilizer mixture (Grade 34 : 17 : 0, with an average weight of 2.7 g per briquette) (Photo 8).

Photo 8. Urea-DAP briquettes prepared with Kranti Briquetter

Inclusion of potash in Urea-DAP briquettes

Potash fertilizers could be included in Urea-DAP briquettes for the supply of K along with N and P. This will increase size of the briquette. So we will have to work out the proportion of N:P:K in the briquette and change the pocketed rollers accordingly.

Most of our Indian soils are medium to high in potassium content. As per soil tests apply potash fertilizer, if necessary through muriate of potash in paired-rows with organic manures. Otherwise, it is not necessary to apply potash, thus reducing the cost of cultivation of small farmers.

Chapter 4

Manufacture of Urea-DAP Briquettes : An Agribusiness

4.1 Introduction

For making rice cultivation profitable in Maharashtra and other States of India, there is a definite need to develop an agrotechnology for small farmers. The agrotechnology for the farmers of 21st century should be simple, less expensive, should maintain soil productivity and also increase grain and straw yields of crops. Considering these needs, Dr. Savant from Pune, developed integrated agrotechnology for transplanted rice, drilled (upland) rice, wheat, sorghum and pearl millet (Savant, 2007). A well planned scheme can produce Urea-DAP briquettes from Urea and DAP fertilizers in the desired ratio of N and P, as per the requirement of a particular region. Also, they can be marketed at a price affordable to small farmers. In short, this energy efficient production strategy at village level will enable the targeting of proper Urea-DAP fertilizer to a specific market at a price which is within the reach of farmers. This is possible because it helps to eliminate the extra bagging, handling and transport costs that would otherwise be required when the same material is produced on a large scale in a factory and then distributed to different rice growing regions of country (Savant and Stangel, 1998). There is an increasing demand for Urea-DAP briquettes required for this agrotechnology by the farmers. For this purpose co-operative societies and youth clubs from rural areas can take up the work of manufacturing and marketing Urea-DAP briquettes in villages. This will be a very profitable business as well as service to the farmers providing guidance for the efficient use of briquettes for different crops.

4.2 Features of the agribusiness

a. Purchase of Urea and DAP fertilizers from the market.

b. Blending of Urea-DAP fertilizers in the proportion of 60:40.

c. Using Kranti Briquetter machine make Urea-DAP briquettes weighing 27 g per 10 briquettes (Photo 8).

d. Selling of briquettes and providing guidelines to the farmers regarding use for integrated (*Charsutri* i.e. four practices / steps) agrotechnologies for transplanted rice, drilled paddy, wheat, sorghum and pearl millet etc.

4.3 Space (land) requirements for agribusiness

- Self owned or rented space, minimum of 15 x 15 m is required for manufacture and sell the briquettes (Fig. 6).
- The four rooms, floored with tiles, should be closed from all sides for protection from rains.

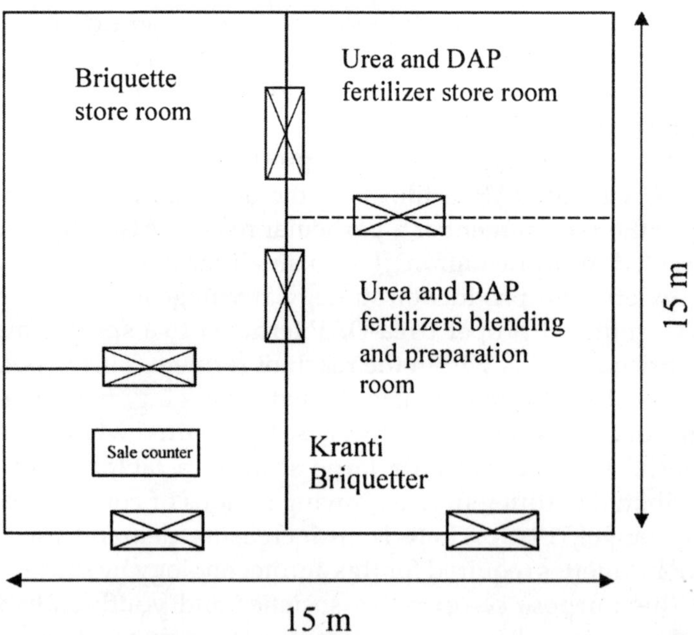

Fig. 6. Map of the space (land) for manufacturing and selling of Urea-DAP briquettes

4.4 Customer

- Large number of small farmers from Maharashtra and all over India can use briquettes for their crops.

4.5 Information about Kranti Briquetter used for manufacture of Urea-DAP briquettes

Technical details (Photo 9)

Photo 9. Kranti Briquetter Machine

- Electric motor of 5 H.P. with gear box.
- Rollers of briquetter and other stainless steel parts with plastic/nylon covers which come in contact with fertilizers.
- Rollers of briquetter should be of good quality and fixed in strong and stout bearings.
- Weight of Kranti Briquetter is about 375-400 kg.
- Size : Kranti Briquetter has length 160 cm x breadth 70 cm x height 160 cm.
- The briquetter requires 3 phase, 440 volts electric current at 50 cycles per second.

Important features

- Minimum maintenance and oiling.
- It is possible to prepare briquettes in different sizes and weights without using a binder.
- Young and educated rural people can get employment to earn money.

4.6 Profit and loss in urea-DAP briquettes agribusiness :

Profit and loss obtained in this business are given in Table 1 (example only).

4.7 Results :

a. Smart and intelligent people may please note that this is just an example of giving profit and loss in this agribusiness.

b. The person preparing briquettes gets 25% subsidy on purchase of Kranti Briquetter and sells to the farmers at Rs.12 to 13 kg^{-1} (as on year 2006). If 45 tonnes of briquettes are sold every year, there will be a loss in the business in the first year. The manager has to be happy with his salary. If he works sincerely and carefully, it is possible to get a profit from 2nd year onwards and repay the loan.

c. Since fertilizers are used efficiently in the form of briquettes no ill effects on the environment are observed. Although, fertilizers are applied up to 60% of recommended doses (40% less) still there is increase in crop yields due to enhanced fertilizer use efficiency.

   Thus, there is a possibility of increasing demand for briquettes for agricultural crops / fruit crops (Savant, 2007).

d. Fertilizer selling Co-operative Societies, Krishi Vigyan Kendras etc. run by agricultural graduates or unemployed youths / farmers can manufacture briquettes as an additional business.

Table 1. Profit and loss in Urea-DAP briquettes agribusiness for the first three years (as on year 2006)

Expenses	1st year	2nd year	3rd year
Permanent expenses			
Rent of the centre	12,000	12,000	12,000
Kranti Briquetter (25% subsidy) (Urea-DAP briquetter manufacturing machine)	1,00,000	-	-
Weighing machine	5,000	-	-
Bag packing electric machine	2,000	-	-
Total Rs.	1,19,000	12,000	12,000
Expenses on raw material			
(45 t briquettes per year)			
Urea (60%) (Rs. 4800/t) - 27 t	1,29,600	1,29,600	1,29,600
DAP (40%) (Rs. 9000/t) - 18 t	1,62,000	1,62,000	1,62,000
Bags 25 (25 kg / bag)	1,300	1,300	1,300
Electricity bill (100 x 240 days)	24,000	24,000	24,000
Total Rs.	3,16,900	3,16,900	3,16,900
Labour and other expenses			
Manager (Rs.5000 x 12 months)	60,000	62,000	64,000
Labour (Rs. 2000 x 2 x 12 months)	48,000	49,000	50,000
Maintenance charges and others (stationary, postage etc.)	15,000	15,000	15,000
Total Rs.	1,23,000	1,26,000	1,29,000

Interest, insurance expenses			
Interest@ 10%	11,900	1200	1200
Current expenses (13%)	41,200	41,200	41,200
Insurance @ 10%	11,900	1200	1200
Total Rs.	65,000	43,600	43,600
Total expenses	6,23,900	4,98,500	5,01,500
Receipts (45 t briquettes per year)			
Sale price / **Total receipts**		**Profit Rs.**	
	1st year	**2nd year**	**3rd year**
Rs. 12 per kg / **Rs. 540000**	-83,900	+ 41,500	+ 38,500
Rs. 13 per kg / **Rs. 585000**	-38,900	+ 86,500	+ 83,500

4.8 Guidelines for manufacture of Urea-DAP briquettes

Manufacture of Urea-DAP briquettes and making these available to the farmers in time at reasonable rates (Savant, 2007).

1. The proportion of Urea-DAP should be 60:40 so that the grade of Urea-DAP briquettes is 34:17:0 of N: P_2O_5 : K_2O, respectively.
2. Average weight of 10 Urea-DAP briquettes would be 27 grams. The briquettes should remain stable and intact during handling / storage.
3. In every polythene (suitable thickness) lining bags only 5, 10 or 25 kg of Urea-DAP briquettes should be placed / stored. Briquette bags should be sealed and carefully handled. The 50 kg bags should not be used, as in such bags the briquettes are likely to be broken into pieces during handling and loss to the farmers.
4. Broken briquettes could be again used to manufacture new briquettes. Broken briquettes should not be placed in the bags for use in the field.

Chapter **5**

Use of Urea-DAP Briquettes for Crops

There is a need for adopting Integrated (*Charsutri*) Organic–Inorganic Farming for increasing per hectare crop productivity and to sustain soil health. For this purpose, it is required to use well decomposed organic manure along with a good quality Urea-DAP (*Vikram*-2) briquettes by farmers as result of increase in benefit : cost ratios. In Maharashtra and other States of India (except sandy soils of Punjab and Haryana), the reasons for using Urea-DAP briquettes for the crops are discussed as below.

5.1 Advantages to the farmers

- Urea-DAP briquettes are manufactured with Kranti Briquetter (which looks like home flour mill) using prilled urea and DAP fertilizers. These briquettes can be easily made available to the farmers in the local fertilizer market (10 to 15 percent higher price than the actual cost of production) with the help of Agriculture Department of the State Government.
- Dr. Savant carried out research work for more than 30 years and developed Integrated Organic-Inorganic Farming Agrotechnology (*Charsutri*) for transplanted rice, drilled paddy, wheat, sorghum and pearl millet (Savant, 2007). There is increase in profit at reduced risk due to increase in benefit : cost ratio. These agrotechnologies are simple, efficient, less expensive and developed for the small farmers. Now there is an urgent need for development of agrotechnologies for sugarcane, vegetables and fruit crops

by regional Agricultural Universities.

- Due to this agrotechnology there will be 40 percent saving on recommended fertilizer doses.
- Due to deep placement of Urea-DAP briquettes in soil, there is less problem of weeds. Therefore, expenditure on weeding will be reduced up to 50 percent.
- Any organic manure, biofertilizer, green manuring, etc. could be used with Urea-DAP briquettes e.g. *gliricidia* leaves along with briquettes is a simple method for limited (only 3 t ha^{-1}) and efficient use of green manuring. Similarly growth of blue green algae in rice fields brings about fixation of nitrogen in soil.

Expected benefits

- Briquettes could also be used with drip irrigation, thus saving both water and fertilizers.

Farmers response

- The training of farmers for deep placement of briquettes will favour rapid adoption of this agrotechnology.

Solution for the above

- An applicator has been developed for deep placement of Urea-DAP briquettes for small farmers for using integrated (*Charsutri*) crop cultivation agrotechnology. Nowthe farmers should be given training regarding integrated (*Charsutri*) transplanting, sowing of crops and deep placement of briquettes by the applicator. Similarly, demonstrations need to be provided for when, how much deep and how many briquettes to be placed in the field.
- Initially Agriculture Officer should organise training for the farmers particularly for deep placement of briquettes in soil. With practice, the farmers can apply the briquettes efficiently and easily.

5.2 Advantages of using briquettes in respect of crops

Deep placement of Urea-DAP briquettes in the root zone of

soil supplies balanced proportion of nitrogen and phosphorus (2 N : 1 P_2O_5) to the plants whenever required for a period of one to two months. The nutrient use efficiency from the Urea-DAP briquettes is 80 to 90 percent (Savant, 2001a). If required, potash could be applied as per the soil testing by mixing it with organic matter in paired-rows only. The advantages of such a practice are :

- Crop growth increases resulting in higher yields.
- Weed problem diminishes in the initial crop growth stages and therefore, less expenses on weeding.
- Due to use of well decomposed organic manure along with briquettes (Integrated Organic-Inorganic Farming), the crop quality is excellent and better soil fertility is maintained.
- The ill effects of conventional use of chemical fertilizers alone due to fertilizer application on the crop leaves could be avoided in deep placement of fertilizer briquettes in soil.
- Expenditure on cost of cultivation of crop is reduced. If one rupee is spent more on use of briquettes, 5-10 rupees profit could be obtained in the same season. Thus, benefit : cost ratio is increased. There is increase in profit at reduced risk (Savant and Stangel, 1998).

Expected benefits

Briquettes prepared from usual urea, DAP and other fertilizers (instead of urea, urea polymers and instead of DAP, polyphosphates etc.) when used for fruit trees the nutrients can be made available for 3-6 months. For this purpose, Agricultural Universities need to develop the technologies for fruit crops.

5.3 Advantages of using Urea-DAP briquettes in respect of soils

- After deep placement in the soil, the briquettes dissolve slowly and the nutrients remain in the soil at a definite place and generates concentration gradients that facilitates

N and P movement as a diffusion flow. Colloidal (complex) particles in the soil adsorb plant nutrients and make them slowly available to the plant roots (Savant and Stangel, 1998).

- Among the four rice hills of the two lines of integrated (*Charsutri*) transplanted / drilled paddy at the site of deep placement of Urea-DAP briquette in the soil, the fertilizer nutrients, particularly phosphorus availability is improved which helps in meeting the requirement of phosphorus by the crop.
- Plant nutrients, particularly soluble phosphorus fixation is reduced thereby increasing its availability to the crop.
- Phosphorus adsorbed on colloidal (clay sorbed) particles of the soils does not leach down (up to 40 percent) along with rainwater (Savant *et al.*, 1997 a).
- Use of briquettes for increasing rice yields in khar land (coastal land) is possible.
- The N_2 fixation by algae in paddy fields is not inhibited as there is little or no fertilizer nitrogen in the floodwater (Roger *et al.*, 1980)
- Unutilized nutrients by the crops, particularly phosphorus remains in the soil which may become available to the crop in the next season.

Important experience

Heavily leached soils (sandy soils of Punjab) or soils with lower CEC (< 10 c mols kg^{-1}) are not suitable for using Urea-DAP briquettes because of losses of plant nutrients through leaching.

Expected benefit

- If micronutrient deficiency is observed in the crops, the same can be provided while manufacturing the briquettes. For this purpose, Agricultural Universities need to take-up research and develop the technology.

5.4 Benefits to the atmosphere

- Due to deep placement of briquettes, ill effects of conventional use of chemical fertilizers, in the atmosphere, soil and ground water are minimized (Daftardar, 2002).

5.5 Benefits to the people

- Due to use of Urea-DAP briquettes rural local people (e.g. Self Help Groups) will get work of preparing briquettes. Labourers working in the field will get work of deep placement of briquettes in the season. Similarly, some people will get work of manufacturing farm implements such as paired-row seed-drill, briquette applicator etc. Thus, even at village level employment is generated in the production and marketing of briquettes.

5.6 Benefits to the State Government

- As expected, the area under cultivation and yield targets of crops per hectare can be achieved. Because of co-operation and help from the farmers, expected crop yields can be achieved, thus ensuring food security.
- The shortage of fertilizers in the crop seasons could be minimized or eliminated, if the State Government plans to manufacture briquettes well in advance. Thus, there are several advantages of using Urea-DAP briquettes for the crops. In case farmers, Agricultural Universities, Department of Agriculture of State Governments and Agricultural NGOs use Urea-DAP briquettes efficiently, the farming business could become profitable.

Fig. 1 Pathways of fertilizer N losses from submerged paddy field

Chapter 6

Spatial Availability of N and P from Deep-placed Urea-DAP Briquettes in Soil

In the intensively cultivated areas of many developing Asian countries, where rice is the staple food and population growth rate is high, declining or stagnant rice productivity is a serious problem threatening the future of food security of the regions (Pingali, 1993). The tropical rice soils are invariably low in organic matter and available N, leading to a widespread response to the application of fertilizer nitrogen. Despite impressive growth in the use of nitrogen fertilizers, particularly urea, nitrogen use efficiency has remained low (Modgal *et al.*, 1995). Poor utilization of fertilizer N by rice is thought to be due to losses from the soil-plant system through ammonia volatilization, denitrification, run off and leaching. Unlike the Green Revolution in rice and other food grain crops in the 1970s and early 1980s, increasing fertilizer use alone may not help to increase the crop yields during the 21st century. This is because conventional fertilizers and their prevailing management practices are not efficient. Therefore, yield increases due to further increase in fertilizer use will be overshadowed by poor economic efficiency, adversely affecting the soil productivity of paddy fields, and, more importantly, aggravating the environmental concerns (De Datta, 1981, 1994; Savant, 1995). For addressing this complex problem, in addition to new ice plant type, farmers need appropriate form of fertilizers for sustainable production of rice and other crops (Savant, 1995).

6.1 Transformations of deep-placed N P briquettes in soil

The soil environment at the placement depth of N P briquettes can play a marked role in influencing transport and transformations of deep-placed N. The physical, chemical and microbiological properties of the rice soil at the placement depth are highly complex, heterogeneous and dynamic (IRRI, 1978 b; Ponnamperuma, 1972).

Research has been done on transport, transformations and availability of nitrogen from deep-placed urea / USG in soil. However, no research has been done on transformations and availability of N and P from Urea-DAP briquettes in soil. The transformations and availability of N from deep-placed Urea-DAP briquettes in soil is similar to that of USG-N / fertilizer N. Regarding the P availability from deep-placed Urea-DAP briquettes in soil, the work has been already done on DAP tagged with ^{32}P. Therefore, for transformations and availability of N from deep-placed Urea-DAP briquettes in soil, the examples and diagrams have been used for fertilizer N / urea / USG.

Considerable spatial variations in soil bulk density and volumetric water content are common in a submerged soil profile (Savant and De Datta, 1980). For 1 to 3 weeks after rice transplanting the soil at the briquette placement depth is generally in reduced condition, but as roots develop, the rhizosphere may become somewhat oxidized because of diffusion of O_2 from rice roots (Armstrong, 1971). However, conditions favourable for denitrification may also develop in some parts of the same rhizosphere, and presumably due to root exudates containing soluble organic substances and dead cell of roots (Kimuta *et al.*, 1979; Russel, 1977). In other words, the rice rhizosphere is considered another site for simultaneous nitrification-denitrification(Reddy and Patrick, 1986; Savant and De Datta, 1982). These temporal changes in the electrochemical properties of soil environment at the placement depths assume importance (a) Nitrogen is likely to be placed at different depths and at different unsaturated to submergence times (during the first few days after planting) and (b) the fate and performance of deep-placed N are compared in different soils. The transport and transformations of urea-N at a typical deep placement site in soil is schematically shown in Fig. 7 (Savant and De Datta, 1980).

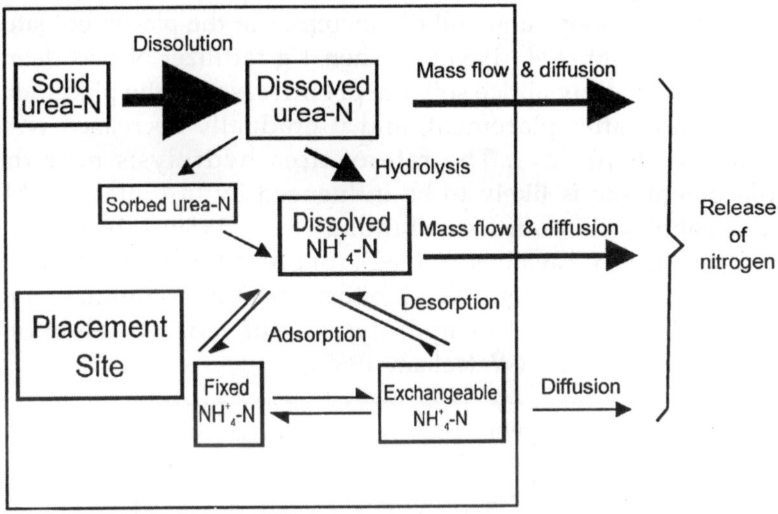

Fig. 7. Schematic representation of uera-N transport and transformations at the placement site in soil

The deep-placed N briquettes dissolve generally within an hour after their placement producing very high localized concentrations of urea at and very near the placement sites. Most of the deep-placed urea in this reaction zone stays in solution, but some of it maybe weekly absorbed by the reduced soil complex (Savant *et al.*, 1982). In coarse textured soils, percolating water causes considerable downward convective transport of urea briquettes from these placement sites (Vlek *et al.*, 1980). Upward movement of urea leading to its entry into the floodwater can occur when proper care is not taken to close the holes at the placement site immediately after deep placement of N briquettes. Craswell *et al.* (1981) found up to 41 ppm urea-N + ammonium-N in the floodwater immediately after deep placement (87 kg N ha^{-1}, dry season), presumably because the dissolved urea moved upward in the water-filled holes left in the soil above the placement sites. With machine deep placement of N, failure to properly close the furrows has been a major problem and has resulted in high amounts of urea-N in floodwater (Khan, 1984).

As enzyme catalysed hydrolysis of urea proceeds in the soil, alkalinity develops and soil pH increases at the placement sites (Fig. 8) (Savant *et al.*, 1983 b). When 1 g fertilizer N was deep-placed in an alluvial rice soil, the pH increased at the placement site 4 days after placement, and it gradually decreased with time up to 10 days. The rate of urea hydrolysis near the placement site is likely to be influenced by (a) unfavourable anaerobic soil conditions (Lindau *et al.*, 1989; Pulford and Tabatabai,1988; Savant *et al.*,1982), (b) prevailing high alkalinity (pH 9), (c) noncompetitive inhibitory effect of accumulation of ammonium (product of hydrolysis), and / or (d) inhibitory effect of urea itself (Wetselaar, 1985).

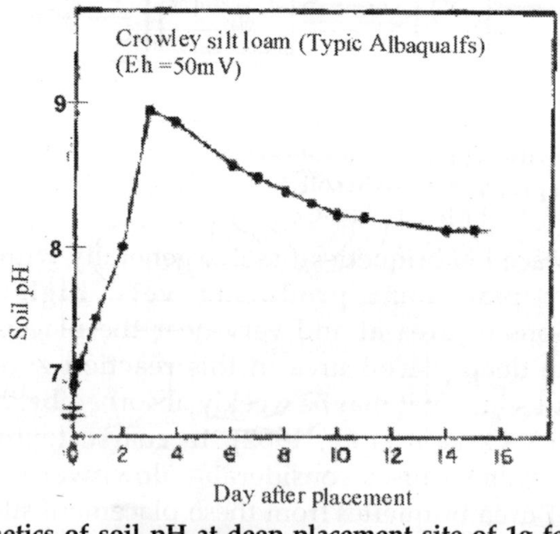

Fig. 8. Kinetics of soil pH at deep placement site of 1g fertilizer N in reduced Crowley soil

The rate of urea hydrolysis at the deep placement sites of USG / briquettes as influenced by submergence time and puddling is not adequately reported in the literature. Recently Eriksen and Kjeldby (1987) reported zero-order kinetics of urea hydrolysis when fertilizer N was deep-placed (10 cm) in two puddled and flooded soils of Norway at 30^0 C incubation temperature. Apparently deep-placed urea as USG is normally hydrolysed in 3 to 7 days after its placement in submerged soils (Craswell and Vlek, 1979; Eriksen and Kjeldby,1987; Vyas and Mistry, 1985).

The ammonium formed as a product of hydrolysis and stable in reduced soil environment, tends to accumulate at the placement sites in three main forms :

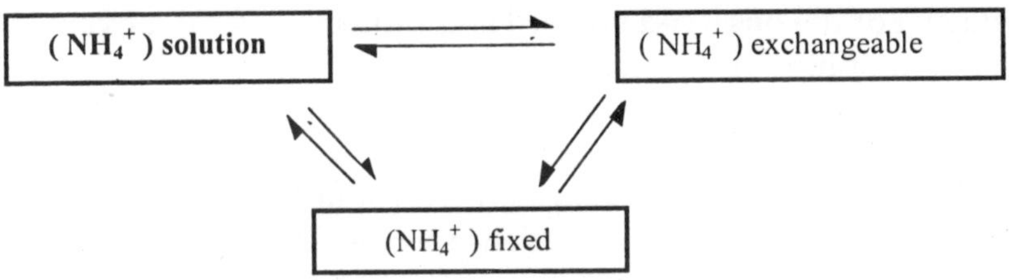

The relative proportion of these forms at placement site varies and mainly depends on the nature and amount of soil colloids. For a typical rice soil, the order of their amounts may be written as follows :

$$(NH_4^+) \text{ exchangeable} > (NH_4^+) \text{ solution} > (NH_4^+) \text{ Fixed}$$

With submergence time, the ammonium is transported mainly through diffusion and interaction with soil. Eventually its fixation by clays and / or microbial immobilization with time may increase. Steep concentration gradients of ammonium exist in the proximity of the placement sites. Figure 2 (Chapter 1) shows in *situ* spatial distribution patterns of KCl-extractable ammonium after deep placement of 2 g USG at about 10 cm depth, in submerged Maahas clay with and without transplanted rice (IR 42).

For a nonpercolating paddy soil with uniform porosity and bulk density, ammonium movement away from the USG placement site is slow which is mainly due to a diffusion-controlled ion exchange process (Savant and De Datta, 1979; 1980). The ammonium concentration gradients seem to exist until the plant absorption occurs (Savant and De Datta, 1980). Similar restricted movement of deep-point placed N as urea in submerged rice soils has been reported by several researchers (Panda and Patnaik, 1989 ; Panda *et al.*, 1988; Rao, 1983). Although movement of ammonium at the placement site of deep-placed USG-N is mainly due to diffusion, at times it could also be due to convection (mainly with downward percolating water). For example, Vlek *et al.* (1980) under controlled conditions observed significant leaching of ^{15}N when USG was deep-placed in silt loam soil and percolation increased from 4.4 mm to 18.3 mm day^{-1}. For the observed upward and probably lateral movement of ammonium diffusion could be responsible. By and large, the order of overall movement of ammonium after deep placement of USG-N in typical puddle submerged rice soil is : downward > lateral > upward (Savant and De Datta,1979; Savant *et al.*, 1982). There is no direct evidence in the literature to suggest concurrent nitrification / denitrification occurring near the placement sites. Nitrates were not detected in the reduced soil solution (collected by porous ceramic cups) at 10-15 cm soil depth after placement of USG in planted Maahas clay (Savant, 1979). However, nitrification seems possible, and some denitrification can, therefore, occur in the rice rhizosphere when rice roots reach the placement sites and start absorbing deep-placed USG-N. In soil solution collected using porous ceramic candles, it was reported that a nitrate nitrogen concentration was raised to 12.5 mg L^{-1} 4 days after deep placement of 2 g USG in transplanted lowland rice soil. In short, denitrification of deep-placed N cannot be completely overruled, but its magnitude is likely to be very small (Cao *et al.* 1984 a and 1984 b). The overall nitrogen transformations and transport processes occurring at and near the deep placement sites of fertilizer N are shown schematically in Fig. 9 (Savant *et al.*, 1983 a).

6.2 Spatial availability of deep-placed N briquettes in soil

Adequate knowledge of spatial reactions such as movement, sorption and chemical or biochemical transformations of deep-placed N as Urea-DAP briquettes is required for better understanding of how and why the new fertilizer works

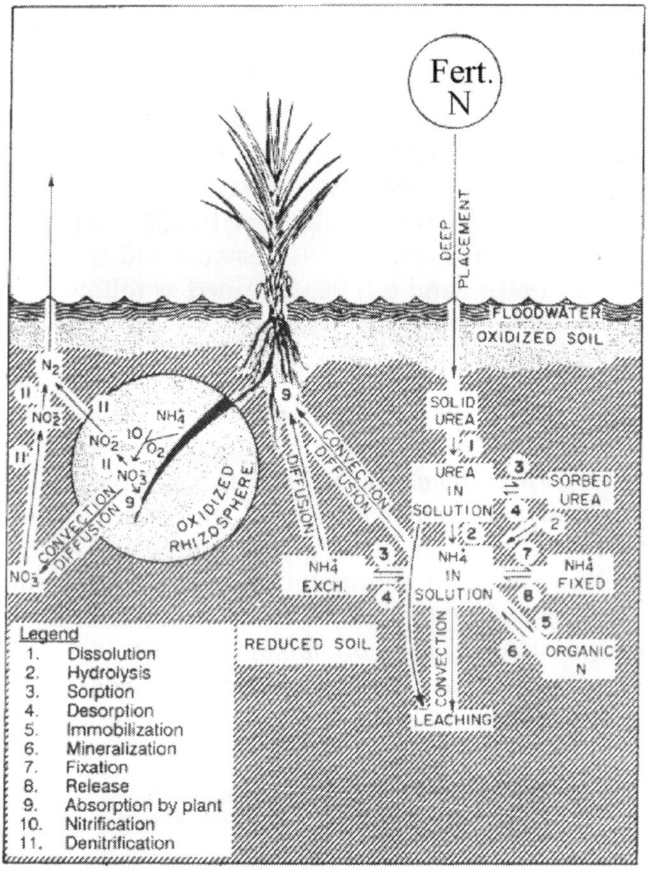

Fig. 9. Overall nitrogen transformations and transport of deep-placed fertilizer N in planted submerged rice soil

efficiently. Availability of conventionally applied or broadcast and incorporated prilled urea-N to rice plants can be adequately estimated from the concentration of KCl-extractable ammonium-N in a representative soil sample collected at random in a given field. But this is not completely true for deep-placed N mainly because point placement of about 1-2 g of urea in the centre of four rice hills at about 10 cm soil depth results in the development of steep concentration gradients of ammonium at its placement sites (Fig. 2, Chapter 1). Therefore, for adequate understanding of plant availability of N briquettes, knowledge of its chemical availability is needed. Although data on KCl-extractable ammonium in soil sample taken at the placement sites will indicate that amount of N available at a given time, the N briquettes may not be readily absorbed by the rice plants because of spatial unavailability. An overall chemical and spatial availability of N briquettes as influenced by soil and plant factors is schematically shown in Fig. 10 (Savant and Stangel, 1985; Savant *et al.*, 1991 a) and can be explained as follows.

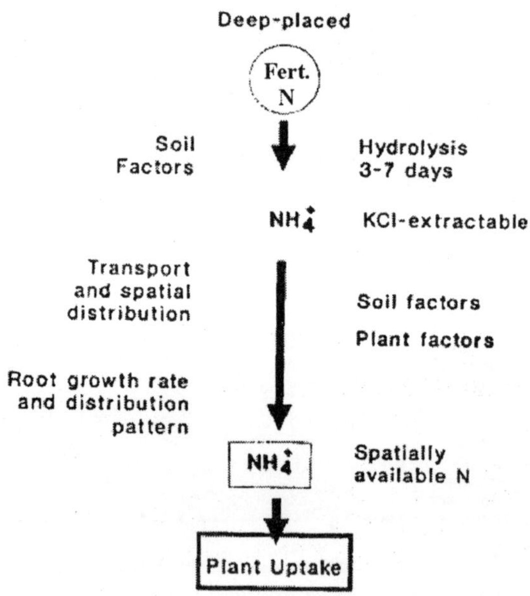

Fig. 10. Spatial availability of deep-placed N in soil

The transport of ammonium from the placement site is slow because it is mainly a diffusion process influenced by ion-exchange. As a result, the spatial concentration gradients of available ammonium tends to exist in a restricted soil volume (Panda and Patnaik, 1989; Rao, 1983; Savant and De Datta, 1979 and 1980). The rice plant may take about 1-2 weeks for its establishment and development of roots and, therefore, there is a lag period before urea-N starts becoming spatially available to the rice plants. During the first 10-20 days after transplanting, Savant *et al.* (1982) observed small ^{15}N plant uptake when 2.6 g USG labeled with ^{15}N were deep-placed at about 10 cm depth immediately after transplanting of rice with 20 x 20 cm spacing in Maahas clay at IRRI in Philippines. In general, increase in the distance between the placement site of urea-N and the rice hills resulted in an increase in the lag period for the initial availability of ammonium to the rice plants,which also seemed to increase with an increase in the depth of urea-N placement (Savant and De Datta, 1980). Tendency of rice roots to avoid high toxic concentrations of ammonium near the placement site of 1 g fertilizer N was observed in root box study (IRRI, 1978 a). As the ammonium concentrations at the placement sites decreased with time, presumably due to continued diffusion and plant uptake, the rice roots might slowly proliferate through the placement sites. At this point, the spatial availability of deep-placed N may reach its peak with time. Thus, the spatial concentration gradients of ammonium existing at the placement sites, control rate and duration of availability of fertilizer N to the rice plant (Savant and De Datta, 1980). This type of spatial availability of N seems to result in an improved relationship between deep-placed N (N source) and the rice plant (sink) as compared with that between traditionally applied prilled urea and rice plant (Chen *et al.*, 1983, Katyal *et al.*, 1984).

From the above discussion, it is apparent that N briquettes *per se* is not a slow release nitrogen fertilizer.

6.3 Spatial availability of deep-placed P in soil

A greenhouse experiment tested the hypothesis that establishing a shorter distance between deep point-placed water soluble phosphorus (P) and transplanted rice hills may help rice plant roots to intercept and recover fertilizer P earlier. The Urea-DAP briquettes tagged with ^{32}P were deep-placed in transplanted rice (Fig.11) (Savant *et al.*, 1993 ; 1994). The soil used was Typic Ustochrepts. The soil was flooded with water for two weeks and then puddled prior to transplanting.

Nitrogen and phosphorus were applied at 58 kg N ha^{-1} and 12 kg P ha^{-1}. Estimated distances between the deep point-placed ^{32}P and rice hills were about 14.1 and 10.6 cm for regular (20 x 20 cm) and modified (15 x 15 cm) spacings, respectively (Fig. 11). As a standard for comparison, ^{32}P-tagged triple superphosphate (TSP) with a specific activity 74 MBq ^{32}P g^{-1} was broadcasted and incorporated before transplanting, and N was applied as prilled urea in two splits (two-third basally broadcasted, and one-third topdressed at the panicle initiation stage). Four hills with 2-3 weeks old rice (IR 36) seedlings $hill^{-1}$ were transplanted in each box and treatments replicated three times under continuously flooded soil conditions.

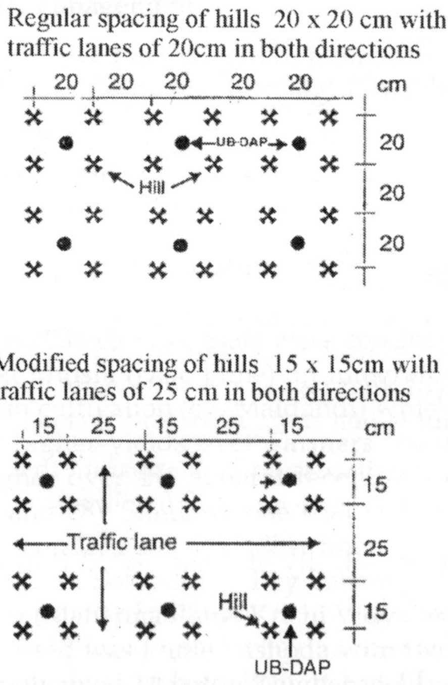

Fig. 11. Schematic sketches of regular (20 x 20 cm) and modified (15 x 15 cm) rice transplanting spacings

The results suggested that basally broadcasted and incorporated soluble P fertilizers in soil are prone to surface run off losses under rainfed, lowland field conditions. The experiments showed practically no ^{32}P in the floodwater when Urea-DAP briquettes were deep-placed. For the incorporated TSP, P uptake reached near plateau 15 days after transplanting (Fig.12). For the deep point-placed P in regular spacing, radioactivity in the youngest leaves was not detected until about 30 and 20 days after transplanting (DAT). Thus, P diffusion alone is not sufficient for the onset of its uptake by rice roots. On the other hand, the onset of P uptake in the youngest leaves was observed about 10 days earlier for the modified spacing (Fig. 12). These and other results suggest that modified spacing can reduce the lag period of spatial nonavailability of deep-placed Urea-DAP briquettes and help to explain improved agronomic performance reported in transplanted rice on the farmers' fields.

Due to phosphate fixation by soil, all the DAP-P stays in the subsoil and diffuses only a few centimeters away from its placement sites. In a greenhouse study without rice plants, ^{32}P as 2.7 g Urea-DAP (N:P = 4:1) deep-placed at about 10 cm soil depth moved 3 cm away from its placement site in about 20 days and up to cm in 30 days after placement (Savant *et al.*, 1997 c).The problem of spatial nonavailability of DAP-P to the rice plants, due to the combined effect of fixation and very restricted movement of phosphate in soil, could be overcome by reducing the distance between the Urea-DAP briquettes and the rice plants (Fig. 12) (Savant *et al.*, 1997 c).

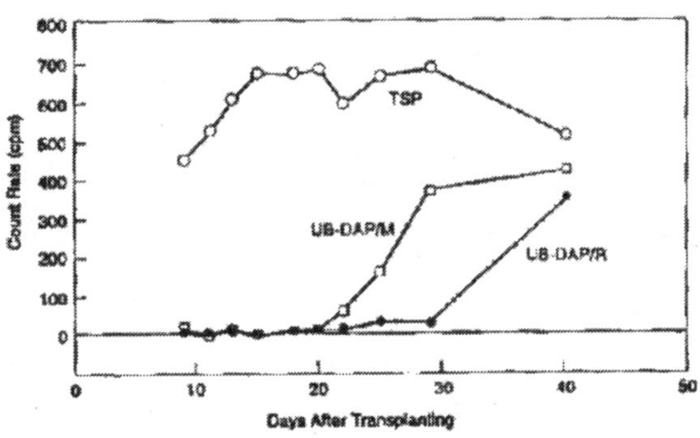

Fig. 12. In situ^{32}P activity in youngest leaves of transplanted rice (IR36) as influenced by basal incroporation of triple superphosphate (TSP) and deep placement of Urea-DAP bariquettes in regular 20 x 20 cm (R) and modified 15 x 15 cm (M) spacings

In a greenhouse experiment when 2.7 g Urea-DAP tagged ^{32}P was deep-placed immediately after transplanting, the shorter distance between its placement site and the rice hills of the modified spacing of 15 x 15 cm apparently enabled the rice roots to intercept DAP-^{32} P about 10 days earlier than the distance of the regular 20 x 20 cm spacing (Fig. 12). This improvement in the spatial availability of fertilizer P could be more important in P-deficient soils. It is thus, apparent that the improved management of Urea-DAP briquettes helps the rice plants to make efficient use of applied nutrients by controlling their rate and duration of bioavailability while reducing the losses of the nutrients (Fig.13). This leads to vigorous growth and healthy plants.

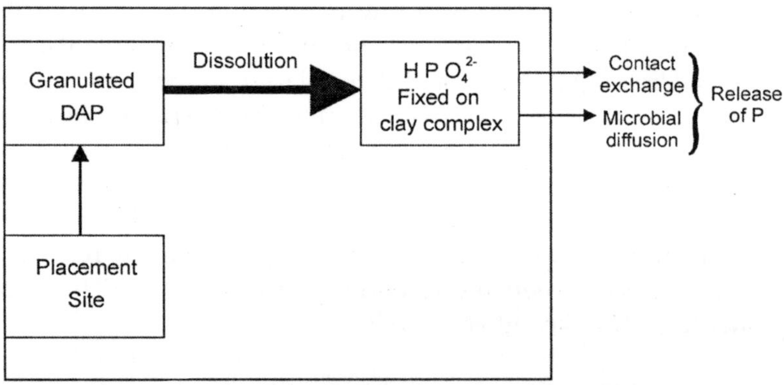

Fig. 13. Schematic diagram of transformations of deep-placed fertilizer P in soil

6.4 Agronomic efficiency of *gliricidia* green leaf N

Soil nitrogen status can have an indirect influence on the agronomic efficiency of deep-placed Urea-DAP briquettes. Transplanted young rice seedlings cannot start absorbing N from deep-placed Urea-DAP briquettes mainly because it is spatially unavailable to them. For a given agroclimate, this lag period may extend up to 4 weeks after transplanting, depending on the depth of Urea-DAP briquettes placement and the transplanting geometry. During the lag period, the young transplanted rice seedlings, therefore, have to largely depend on the soil N status (ammonium-N pool and N supplying capacity). This means that the initial soil ammonium status will determine the initial plant growth stand and can have significant impact on yield (Mengel *et al.*, 1986; Schnier *et al.*, 1988). The very low N status of many soils in the tropics (Kyuma, 1989) may result in overall poor crop growth stand especially with regard to vigour and tillering (Fig. 14) (Savant and Stangel, 1985).

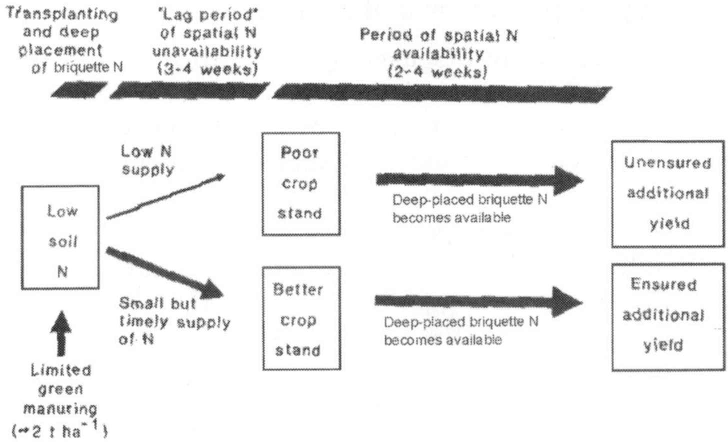

Fig. 14. Effect of low soil N on agronomic efficiency of basal deep-placed Urea-DAP briquettes

This observation, therefore, suggests that very low soil N can limit the agronomic efficiency of basal deep placement of Urea-DAP briquettes. In other words, single basal deep placement of Urea-DAP briquettes alone may not adequately ensure increase in yields in soils with very low N status. Under these soil-N conditions, a small amount of N as prilled urea (¼ of total N dose) or limited green manure may be incorporated in soil before transplanting to meet at least in part the initial N requirement of the rice plants and to provide a reasonable initial crop stand. Incorporation of *gliricidia* leaves (about 3 t ha^{-1}) in soil during puddling operation, supply all micronutrients to the transplanted rice (Dhane *et al.*, 1991). Evolution of gases such as methane after anaerobic decomposition of the limited leaf biomass from flooded rice field can be reduced proportionally, causing less air pollution. Above practice of organic farming helps farmers to sustain soil productivity and to increase rice yield with less pollution (Savant, 2002).

Using a holistic approach, the Urea-DAP briquettes management can be integrated with plant nutrient recycling and limited *gliricidia* green leaf manuring. Small resource-poor farmers can use this nutrient management system for increasing and sustaining rice production in the future.

From the foregoing discussion, application of Urea-DAP briquettes at the time of planting or 10-15 days after sowing has been found beneficial. Spatial availability of N and P from Urea-DAP briquettes in soil to the crop indicates two important findings.

i. Participating farmers in the demonstration trials observed that transplanted rice crop was greener in colour throughout the season, suggesting that N and P were spatially made available to the crop plants.

ii. After harvest of the crop, the total uptake of N and P was always more than that of the normal crop. This also suggests that N and P requirements of the crop were fulfilled.

Therefore, while using N P briquettes in soil, paired-row planting of rice is useful. Similarly, for cereals and pulses paired-row sowing is beneficial.

Chapter 7

Principles in the use of Urea-DAP Briquettes for Crops

7.1 Introduction

Plant growth and its ultimate end results (agronomic yields) are influenced by the total effects of all interactive environmental factors (organic and inorganic) in an open field. (Fig. 15). Therefore, either of these factors in field, eg. use of inorganic fertilizers (Urea-DAP briquettes) alone cannot be considered in isolation. The use of Urea-DAP briquettes need to be considered along with all other remaining environmental factors (air, soil and water etc.) in the field. Therefore, by taking into account crop cultivation practices four (or five) important factors are considered in the so called *Charsutri* or integrated approach (Savant, 2007; Savant and Sonar, 2007 a).

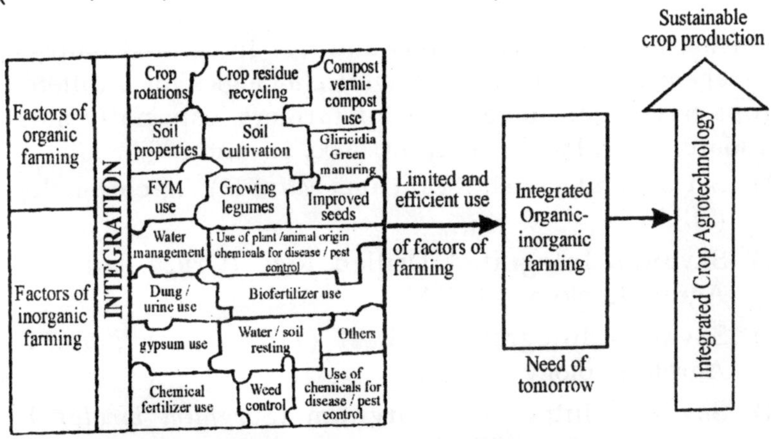

Fig. 15. Cause-effect relationship between factors of integrated organic-inorganic farming and sustainable crop production (different shapes of the factors indicate their integration with others)

7.2 For efficient and economically viable use of briquettes for agricultural crops, the following points should be considered.

- Enough supply of Urea-DAP briquettes are available to the farmers in local markets in every season. Urea-DAP briquettes is the source of N and P supply.
- A well proven agrotechnology should be made available to farmers (only use of briquettes is not enough).
- Use of suitable plant geometry and appropriate plant population of crop with suitable variety.
- When and how briquettes should be made available.
- For transplanting rice crop 15 cm x 15 cm spacing between paired-rows with a walking lane of 25 cm distance, while for drilled rice 15 x 15 cm spacing between paired-rows with a walking lane of 30 cm distance in both horizontal directions are desirable. This is all required for adequate supply of N and less mobile P (than N) to the crop at early growth stages.
- Apply one Urea-DAP briquette for 15cm x 15cm square of 4 hills at a distance of 30 cm in paired-rows for cereal crops for small resource-poor farmers to keep the cost of cultivation minimum.

Considering the above points and five Integrated Nutrient Management System (INMS) agrotechnologies for 5 different crops have been developed for farmers, especially small (resource-poor) farmers (Savant, 2000, Savant, 2007).

(1) Savant's Integrated Rice (*Oryza sativa* L.) Agrotechnology (SIRA)

(2) Savant's Integrated Drilled Rice (*Oryza sativa* L.) Agrotechnology (SIDRA)

(3) Savant's Integrated Wheat (*Triticum aestivum* L.) Agrotechnology (SIWA)

(4) Savant's Integrated Sorghum (*Sorghum bicolor* L.) Agrotechnology (SISA)

(5) Savant's Integrated Pearl Millet (*Pennisetum americanum*) Agrotechnology (SIPMA)

Savant's Integrated Rice Agrotechnology (SIRA)

The following are the four steps or practices in SIRA.

Step 1 : Farmer recycles rice crop residues (hull and straw).

Step 2 : During final puddling and before transplanting, farmer incorporates limited *gliricidia* green manuring (~ 3 t ha^{-1}) in soil using agroforestry approach.

Step 3 : For all varieties, soils and seasons, the farmer uses controlled transplanting using 25 hills m^{-2} area.

Step 4 : Farmer inserts urea-diammonium phosphate (Urea-DAP) briquettes (each briquette weighing 2.7g for 4 rice hills) by hand at 7 to 10 cm soil depth on the same day after controlled transplanting of rice.

For other cereal crops (drilled method) the following are the practices or steps.

Step 1 : Using seed treatment of improved or hybrid crop sowing / dibbling is done in paired-rows (15 x 15 cm spacing) in south-north direction as far as possible.

Step 2 : Efficient and limited use of organic manure in paired-rows only.

Step 3 : Efficient and limited use of Urea-DAP briquettes per 30 cm distance in paired-rows.

Step 4 : Efficient use of available water for irrigation (sprinkler or drip or alternate furrow method of irrigation).

The importance of environmentally sound as well as agronomically efficient practice of recycling rice plant residues is summerised in Fig. 16 (Savant, 2001 a).

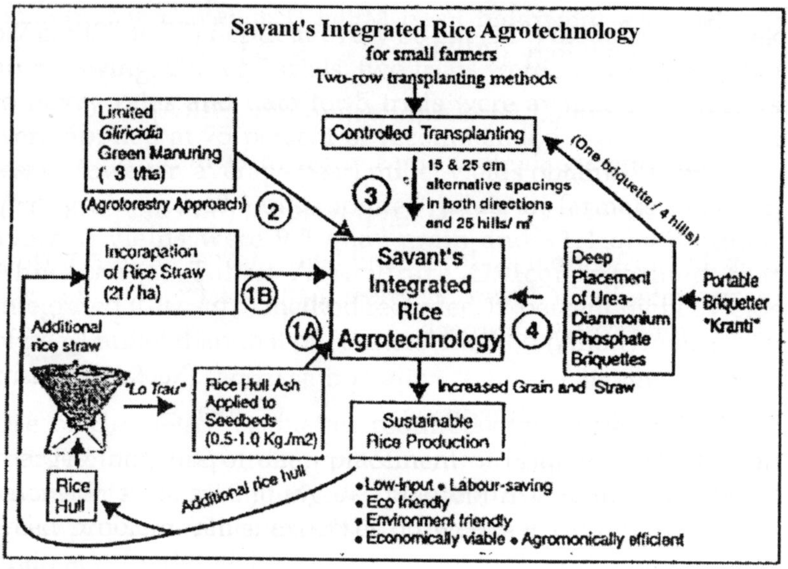

Fig. 16. Savant's Integrated Rice Agrotechnology (SIRA)

Chapter 8

Integrated use of Urea-DAP Briquettes for Transplanted Rice

In 21st century, rice scientists and planners of South and Southeast Asia are facing a challenge of growing about 70% more rice from the decreasing agricultural land without loosing quality of natural resources and the environment. For facing the challenge of food security and sustainable agricultural development, a potential option to the rice planners is to enhance productivity of millions of paddy fields of small farmers in rainfed lowland areas. Because green revolution agrotechnologies are not appropriate for small farmers, they continue to practice traditional rice cultivation and harvest low yields. Millions of these farmers need simple, low-cost input and environmentally benign, Integrated Nutrient Management System (INMS) for enhancing and sustaining rice yields.

The agrotechnology is based on Integrated Nutrient Management System (INMS) approach and it can be defined as "crop cultivation system that integrates efficient use of soil and applied plant nutrients with appropriately modified agronomic and pest management practices, aimed at increasing and sustaining crop production for a long time while conserving the national resources and protecting the environment" (Savant, 2001 a).

Based on Integrated Nutrient Management System for rice, since 1978 research was conducted at International Rice Research Institute, Philippines; International Fertilizer Development Centre, USA; and Agricultural Universities in Maharashtra

(Savant and Stangel, 1995 ; Stangel *et al.*, 1993; Savant and Stangel, 1998). Using holistic approach, Savant's Integrated Rice Agrotechnology (SIRA) has been developed specially for small and resource-poor rice farmers of Maharashtra and other States of India (Savant, 2001 a; 2000).

8.1 SAVANT'S INTEGRATED RICE AGROTECHNOLOGY (SIRA) : MORE RICE WITH LESS POLLUTION

Because of its "Integrated Nutrient Management System" (INMS) principle, SIRA may have potential for use in other States of India and also for developing countries of Asia. The SIRA consists of four agronomic practices that fight pollution (Savant, 2000; Savant and Daftardar, 2000; Savant *et al.*, 2004).

Practice 1. Recycling of Rice Crop Residues (Hulls and Straw)

1 a. Application of rice hull ash (RHA) for growing rice seedlings

Farm women can use rice hull ash stove for domestic cooking (eg. Vietnamese Low Trau). Black grey rice hull ash (RHA) as a source of silicon produced by the stove is saved throughout the year and incorporated into nursery seedbed @ 0.5 to 1.0 kg RHA m^{-2} before sowing of seeds of improved / hybrid cultivars of rice for growing healthy seedlings (Photo 10 a) (Savant and Sawant, 1995; Savant *et al.*, 1997 b; Sistani *et al.*, 1997).

Photo 10 a. Rice seedlings (var. Jaya) grown with and without rice hull ash

Benefits of supplying silicon through RHA to rice seedlings

- The rice seedlings become healthy and tough with well developed root system (Photo 10 b) (Savant and Sawant, 1995; Savant *et al.*, 1997 b; Sistani *et al.*, 1997).

Photo 10 b. Rice seedlings (var. Indrayani) 30 days old grown with and without rice hull ash

- The seedlings develop built-in resistance to protect themselves from stem borer and blast attack in the nursery (Sawant *et al.* 1994; Kumbhar et al., 1995).
- The seedlings can utilize applied P and K more efficiently (Savant and Sawant, 1995).
- The seedlings can reduce toxic effects of Fe, Mn and Al ions in acid soils (Sistani *et al.*, 1997; Savant *et al.*, 1997 b).
- The seedlings have better plant growth and development and, therefore, produce more rice yield (Sawant *et al.*, 1994; Kumbhar and Savant, 2001 a and 2001 b; Powar and Deshpande, 2000).
- The roots of RHA-treated rice seedlings have the ability to oxidize methane (CH_4) generated in submerged rice soil. However, this report warrants more research work to reduce methane coming out of rice fields.

1 b. Recycling of rice crop residues (hulls and straw)

The present agrotechnology produces more straw than the conventional method. Therefore, a farmer can incorporate the increased rice straw (~2 t ha^{-1}) for recycling into the soil. The remaining straw can be used for feeding cattle in dry season. The dry rice straw is applied to soil while preparing the land by ploughing. The dry rice straw is evenly spread @ 2 t $ha^{-1}$and incorporated into the soil during first or second ploughing for land preparation. Subject to availability of farm labour, straw can be cut into two or three pieces and then it can be incorporated into the soil in presence of some water. Both these conditions help quicker incorporation and decomposition of straw.

Benefits of the incorporation of rice straw in soil

- Potassium present in the straw is applied to rice crop (Savant *et al.*, 1997 b). This eliminates the need for applying expensive imported potassic fertilizer. This practice saves money for rice farmer, foreign exchange of a nation and improves fertility of a soil.
- About 2 t ha^{-1} straw could supply 20 to 30 kg K_2O ha^{-1} and 60 to 110 kg Si ha^{-1}. Due to supply of K and Si, the rice crop can protect itself from blast and leaf scald (Kumbhar and Savant, 2001 a; 2001 b; Savant and Sonar, 2007 b).
- Rice straw adds organic matter to the soil.
- Straw helps to increase rice yields (Bulbule *et al.*, 1996).

Practice 2. Limited Use of *Gliricidia* Green Manuring Using Agroforestry Approach

Using *gliricidia* (*Gliricidia sepium* (Jacg.) Steud) green manuring is an old practice. However, the practice suggested in this agrotechnology is more simple and affordable to small farmers.

In this practice a farmer uses limited *gliricidia* green manuring (~3 t ha^{-1} green biomass) in soil using agroforestry approach (Fig. 17). During final puddling and before transplanting, farmer incorporates *gliricidia* green manuring.

Benefits of application of *gliricidia* green manuring

- It requires green biomass of about 3 t ha^{-1} (or 3 - 4 trees for every 100 m^{-2} field) as compared to that of 7 to 10 t ha^{-1} in any conventional or recommended green manuring. Therefore, this limited *gliricidia* green manuring practice is practical, less time consuming and affordable to farmers.
- It does not require a rice field and a growing season for growing green manuring crops like sannhemp, dhaincha etc. Also it does not require seeds and cultivation every season as *gliricidia* is a perennial tree and grown on the bunds of the field.
- *Gliricidia* leaves decompose very rapidly in submerged soil.

Fig. 17. *Gliricidia* trees grown on the bunds (borders) of the field and surrounding the house

Its N (10 to 12 kg ha^{-1}) is timely supplied to rice seedlings during the first 3 weeks after transplanting, that is when the deep-placed N in the briquettes is spatially not available to the seedlings (Savant and Stangel, 1990). This ensures better growth and ultimately higher yield of rice.

- *Gliricidia* as a green manure can supply all the macro and micro nutrients to the rice crop.
- Limited leaf biomass results in limited amount of methane generated in submerged rice soils. This helps less methane pollution as compared to that of recommended dose of 10 t ha^{-1} leaf biomass of *gliricidia.*

Practice 3. Controlled Transplanting of Rice Seedlings

Random and uncontrolled transplanting of seedlings of local varieties of rice is one of the reasons for poor yields of small farmers in the developing countries. In this practice, a farmer follows controlled transplanting of seedlings of an improved rice variety. The RHA-treated 3-4 weeks old healthy seedlings of rice are transplanted at 15 cm and 25 cm alternate hill spacings in both horizontal directions and at 25 hills m^{-2} (Fig. 18) (Savant, 2000; 2001 a).

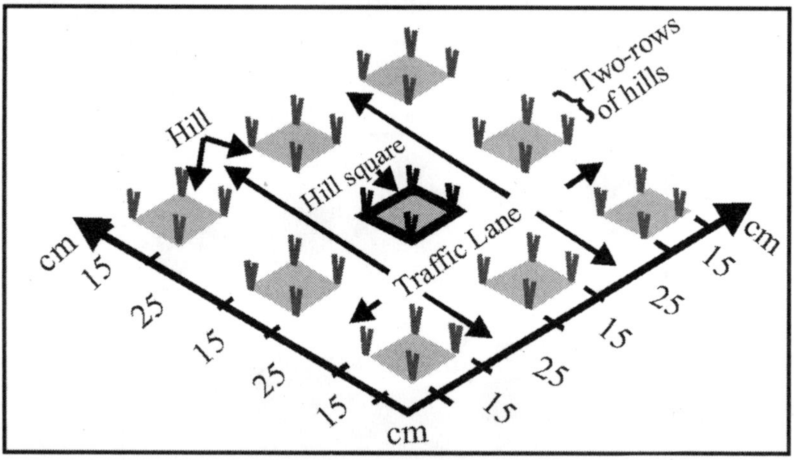

Fig. 18. Controlled transplanting of rice (15 x 15 cm paired-rows with a walking lane of 25 cm)

- To achieve this, recently developed two- row or paired-row transplanted method is followed for formation of 15 x 15 cm hill squares and 25 cm traffic lanes in entire rice field (Figs. 11, 18 and 19) (Savant, 2001 a; Savant, 2006).
- Small farmers can use *Bamboo* Guide Transplanted Method (Fig. 20). and other farmers can use Modified Transplanting Line Method (string having distances of 15 cm and 25 cm alternate hill spacings) (Savant, 2001 a).

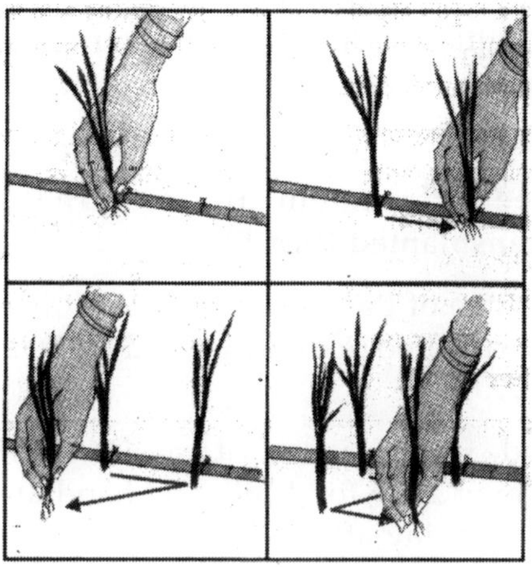

Fig. 19. Transplanting of rice at 15 x 15 cm hill squares

Benefits of controlled transplanting

- It saves seeds (up to 50%) and labour charges for transplanting and harvesting (up to 30%).
- This lower hill population (25 hill m^{-2}) than the higher population of 33 or more hills m^{-2} helps to reduce methane emitted from rice soils for the first month after transplanting.

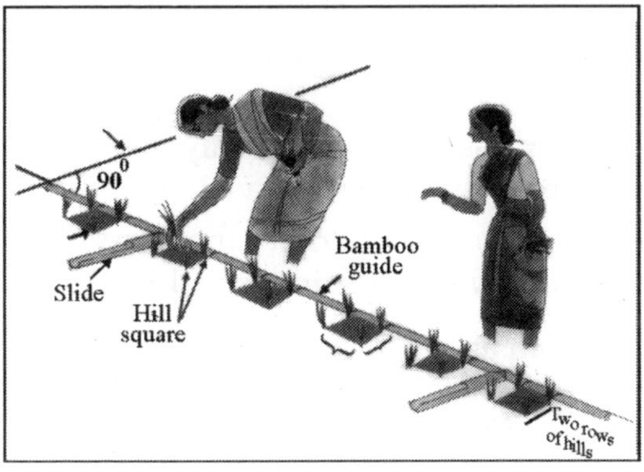

Fig. 20. *Bamboo* guide method of controlled transplanting of rice

- It facilitates efficient use of deep-placed Urea-DAP briquettes into the soil (Practice 4).
- Distances of 25 cm between 15 cm paired- rows of hills are traffic lanes. These lanes are very convenient for workers

to walk while deep- placing the briquettes in soil by hand.

- 15 x15 cm squares, each of 4 hills, in rice field helps workers to easily locate the sites of briquette placement (Practice 4).

Practice 4. Deep Placement of Urea-DAP Briquettes in Transplanted Rice Field

This practice is one of the most efficient fertilizer management methods in transplanted rice (Savant and De Datta, 1982; Savant and Stangel, 1990).

In this method, farmer inserts Urea-DAP briquettes (each briquette weighing 2.7 g for 4 hills) by hand at 7 to 10 cm soil depth on the same day after controlled transplanting (see Fig. 21 A and 21 B).

Benefits of deep placement of briquettes

- As compared to recommended fertilizer dose (100 kg N + 50 kg P_2O_5 + 50 kg K_2O ha^{-1}), farmer saves 40% N and P and 100% K fertilizer and practically eliminates applied N and P losses, thus resulting in nearly no fertilizer related pollution.
- There is less weed problem and, therefore, less weedicide related pollution in the rice field that is fertilized with briquettes.
- This fertilizer use is agronomically efficient (up to 80%) and consistently increases rice yield (average 1 t ha^{-1}).
- This practice has benefit : cost ratios varying from 5 to 10 and is practically risk free in normal rainy seasons.

(**Note :** Farmers can use 10 to 20 kg K_2O ha^{-1} through muriate of potash at final puddling in potassium deficient soils).

8.2 Highlights (1993 to 2007)

- In several field trials, transplanted rice (improved varieties, early to late, and hybrid rice, e.g. Sahyadri) on different soils (Inceptisols, Alfisols and coastal saline soils) of Maharashtra State have responded well to deep-placed Urea-DAP briquettes in wet and summer seasons.

- In more than 1000 field trials, Janaseva Foundation (Silvassa), Pune has demonstrated that tribal rice farmers of Raigad, Thane, Nasik, Nandurbar and Dhule districts can nearly double their rice yields by using Urea-DAP briquettes.
- Government of Maharashtra is providing 25% subsidy for purchasing Kranti Briquetter since the year 2000. Department of Agriculture, Maharashtra State, Agricultural Universities, several NGOs are extending SIRA to the rice farmers of State and the farmers are responding extensively to SIRA.

8.3 Conclusions

- The SIRA is a simple, agronomically efficient, economically attractive and environment friendly.
- The SIRA using hybrid or improved rice variety is a boon to small (resource-poor) rice farmers of Maharashtra and other States of India.
- The SIRA has potential for enhancing rice yields of farmers in Andhra Pradesh, Karnataka, Tamil Nadu, Kerala etc. (SIRA is not suitable for sandy soils of Punjab).

8. 4 Use of Urea-DAP briquettes

The practice of using Urea-DAP briquettes is one of the most efficient fertilizer management methods in transplanted rice (Savant and De Datta, 1982; Savant and Stangel, 1990). Deep placement of Urea-DAP briquettes @ one briquette for every four hills (15cm x 15cm hill square) is done by hand (Savant and Stangel, 1995) or PVC made briquette applicator (Savant *et al.*, 1991 a; 1992) at 7 to 10 cm soil depth on the same day of controlled transplanting (Daftardar *et al.*, 1997a; 1997b; 1997c; Powar and Deshpande, 2000 ; Savant and Chien, 1990; Savant *et al.*, 1993; Talashilkar *et al.*, 2000).

8.4.1 Time of Urea-DAP briquette placement

Urea-DAP briquettes should be inserted 7 to 10 cm (3″ to 4″) deep into the soil immediately after controlled transplanting. The deep placement of briquettes in freshly puddled soil (mud) is done on the same day for two important reasons.

i. The puddled soil condition at transplanting is soft and, therefore, it is the most favourable condition for inserting briquettes 7 to 10 cm deep by hand (Fig. 21 A).

ii. Soft puddled mud (soil) helps to close the holes of the briquette placement sites left by hand immediately and naturally (Fig. 21 B) (Savant and Stangel, 1995 and 1998).

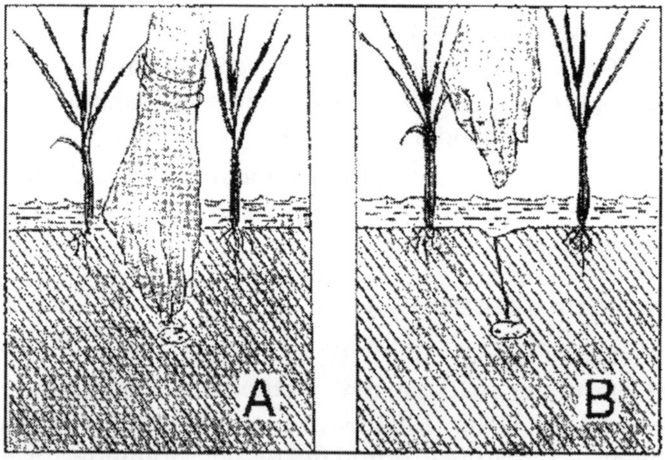

Fig. 21 (A) Least soil resistance for inserting briquettes on the same day of transplanting (B) Holes left at briquette placement sites are naturally closed

IMPORTANT: Any delay in inserting briquettes for a day or more after transplanting leads to settling down of the puddled soil. This increases its bulk density with time. The settled soil even in standing floodwater may pose serious problems in inserting briquettes by hand at 7-10 cm depth. Workers may have a tendency to place briquettes at shallow depth or even at surface of the field which is undesirable (Figs. 22A and 22 B) (Savant, 2001 a).

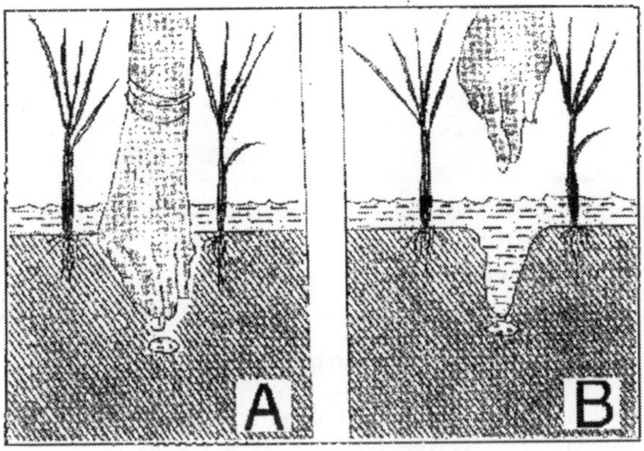

Fig. 22 (A) Problem for inserting briquettes on the same day of transplanting (B) Holes at briquette placement site remaining open

If briquettes are inserted in the settled soil with some force by hand (Fig. 22A), the holes at insertion sites of briquettes may remain open (Fig 22 B). Open holes are undesirable because they permit nutrient losses and downstream pollution. The workers placing briquettes will have to close the holes immediately. This closing of the holes is practically a very difficult operation under the settled soil conditions.

In short, the delayed placement of briquettes may lead to their decreased agronomic efficiency for the rice crop.

8.4.2 Method of briquette placement

1. A worker should tie a plastic bag around the waist and take 1 to 2 kg of briquettes into the bag (Fig. 23 A).
2. In order to avoid moisture problem, the worker must collect 5 to 10 briquettes at a time from the bag with the dry left hand (Fig. 23 B).
3. The worker drops only 1 briquette from the left hand into the right hand (Fig. 23 C).
4. The worker inserts 1 briquette with his right hand 7 to 10 cm deep into the soil at the centre of the 15 cm x 15 cm square of the four hills (Fig. 23 D) (Talashilkar and Savant, 1991 ; Daftardar and Savant, 1995).

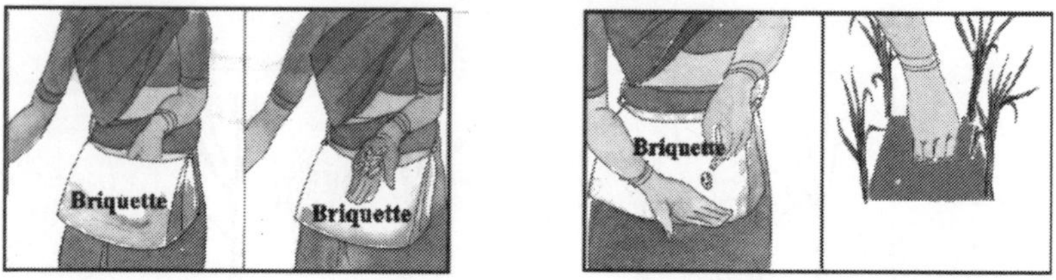

A. Taking briquettes from bag **B. About 10 briquettes in let dry hand** **C. Dropping one briquette from left hand to right hand** **D. Deep placement of briquette per 4 hills**

Fig. 23. Method of application of briquettes

(a) A short worker walks in one traffic lane and inserts briquettes in two paired-rows of hills on either sides (Fig. 24) (Savant, 2001 a).

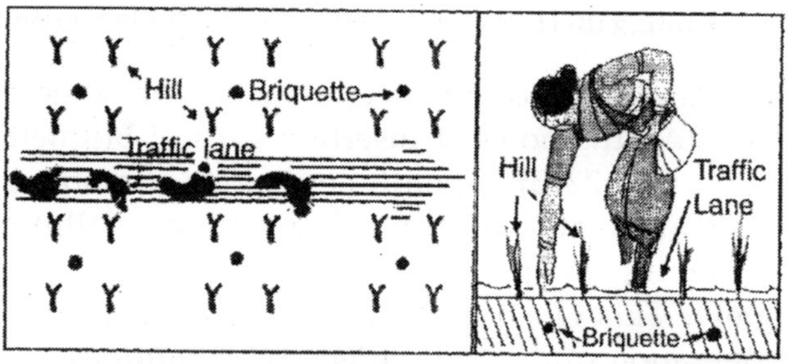

Fig. 24. A short worker inserting briquettes in two paired-rows

(b) A tall worker walks into two traffic lanes, keeping one foot always in one traffic lane and the other foot in the next traffic lane and inserts briquettes in three paired-rows of hills (Fig. 25).

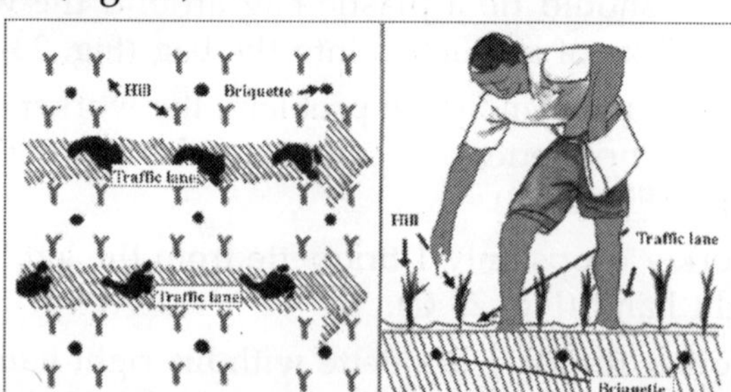

Fig. 25. Deep placement of briquettes in field by tall worker in three paired-rows

- Take fertilizer grade urea-diammonium phosphate mixture in the proportion of 60 : 40 and prepare Urea-DAP briquettes, each weighing 2.7 ± 0.2 g (see Photo 8). Urea-DAP briquette can be prepared with Kranti Briquetter at 200 to 250 kg hr^{-1} (Photo 9). It requires 625 number of briquettes or 1.75 kg briquettes for every 100 m^{-2}.
- Fertilizer dose applied is 58 kg N +29 kg P_2O_5 ha^{-1}.

8.4.3 Main benefits of application of Urea-DAP briquettes

- This is the most efficient method of fertilizer use in transplanted rice.
- Up to 80 percent of applied nitrogen is used for rice crop and the losses are practically minimum. As a result, fertilizer use efficiency is maximum and fertilizer related pollution is almost eliminated (Savant *et al.*, 1982 ; Daftardar *et al.*, 1997 a and 1997 c).
- Weed menace is less due to deep placement of briquettes and, therefore, use of weedicide or weeding expenses can be eliminated (Daftardar *et al.*, 1997a).
- Average rice yield (grain and straw) is increased up to 50 percent or 1 t ha^{-1} (as compared to farmers' method).
- This practice is economically attractive. If a farmer spends Rs. one for Urea-DAP briquette use, he can get in return Rs. 5 to 10 in the same season (Powar and Deshpande, 2000).

8.4.4 Some important facts about the briquette placement

- Briquette placement is affordable in controlled transplanted rice only (Fig. 26).

Fig. 26. A happy worker placing briquettes in controlled transplanted rice field

The deep placement of briquettes is easy, simple and less time consuming job in the controlled transplanted rice field (Fig.26). With some training, even high school boys and girls can do it without any difficulty.

There are 15 cm x 15 cm hill squares distributed through out the controlled transplanted field. Using one or two traffic lanes for walking, workers can easily identify these hill squares, that is briquette placement sites. Without taxing the brain, a worker can easily insert the briquettes in two or three paired-rows of hills (see Figs. 24 and 25).

The number of briquettes placed is automatically controlled by placing one briquette for every four hills (Fig. 23 and Photo 11). This also helps to control the rate of N and P applied per hectare.

In the randomly transplanted rice field, workers are always confused in placing briquettes (Fig. 27). This placement is very difficult and time consuming operation because workers will have to use their own judgement for placing each and every briquette in the field. This operation may be possible in small experimental plots using experienced workers only. However, it is totally impossible for an average worker to place the briquettes in the soil in randomly transplanted rice field.

Fig. 27. A confused worker in randomly transplanted rice field

- Crop stand of randomly transplanted rice fertilized with briquettes will be uneven

Placement of briquettes in the randomly transplanted rice will result in uneven crop stand (Fig. 28) (Savant, 2007). This is because N and P in the briquettes are not evenly supplied to all the seedlings per hills. They are available to nearby, two to three or four hills only, leaving other distant hills away from the briquettes starving for the applied nutrients.

Another disadvantage is that workers cannot control the number of briquettes applied in the randomly transplanted rice. Therefore, N and P applied is not controlled and the farmer cannot save fertilizers and money as well.

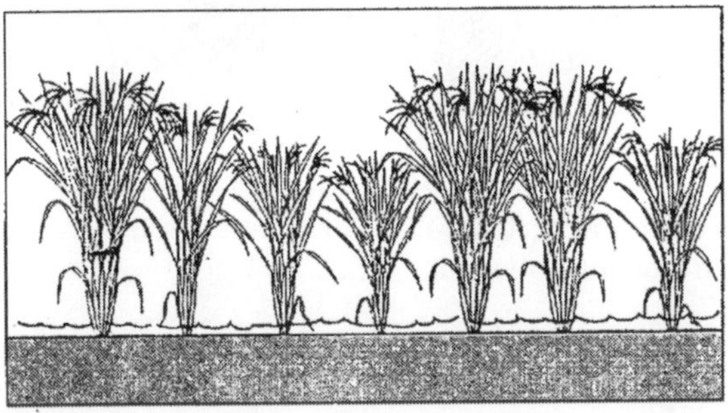

Fig. 28. Uneven crop stand of randomly transplanted rice fertilized with briquettes

Photo 11. Visit of Dr. Savant to rice (cv. Sahyadri) trial on farmers' field at Kolwan, Tah, Mulshi, Dist. Pune

- **Never broadcast Urea-DAP briquettes in rice fields**

Broadcasting of Urea-DAP briquettes in transplanted rice fields is never efficient and can pollute the environment (Fig. 29). They are not meant for broadcasting in transplanted rice. They must be deep-placed for agronomic efficiency.

Fig. 29. Broadcasting of Urea-DAP briquettes is inefficient and causes environment pollution

- **One briquette deep-placed for every four hills**

One Urea-DAP briquette is deep-placed for every four hills square in the controlled transplanted rice. This is the correct application of the briquette (Fig. 30).

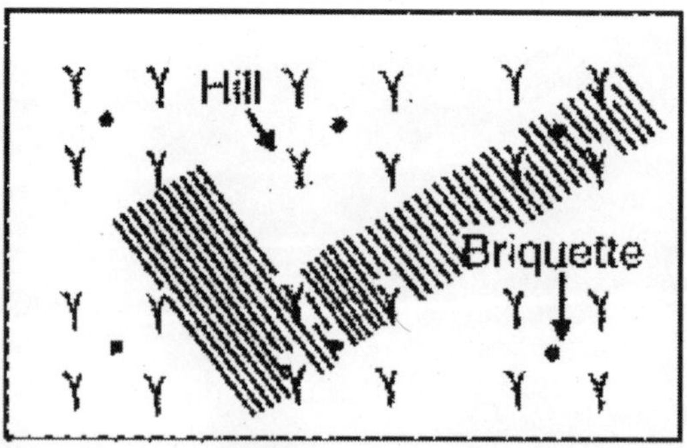

Fig. 30. Correct method of application of briquettes in transplanted rice field

Placing 2 to 4 Urea-DAP briquettes for every two hills or four hills of rice (Fig. 31) is a wrong application of briquette placement. For example, assuming a worker requires one second extra to count two or more briquettes for every four hills having 20 cm x 20 cm spacing (2,50,000 hills ha^{-1} or 62,500 placement sites ha^{-1}), he will need minimum additional 2 man days ha^{-1}. This increase in additional labour cost will discourage the farmers and decrease chances of their accepting this practice (Savant and Stangel, 1995 and 1998). Further the objective of controlling the number of briquettes used and thereby the rate of N and P applied per hectare will be defeated.

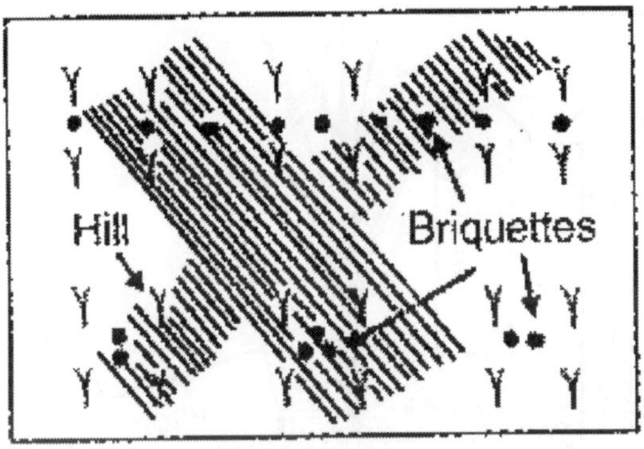

Fig. 31. Wrong method of application of briquettes in rice field

- **Effectiveness of briquettes depends on 7 to 10 cm deep they are placed in the soil**

When briquettes are placed 7 to 10 cm soil depth, fertilizer losses are reduced to less than 5 percent and agronomically fertilizer use efficiency of briquettes is maximum i.e. up to 70-80 percent (see Fig. 24). (Savant and De Datta, 1982; Savant *et al.*,1982; Savant and Stangel, 1998). This application method is environment friendly (Fig. 32).

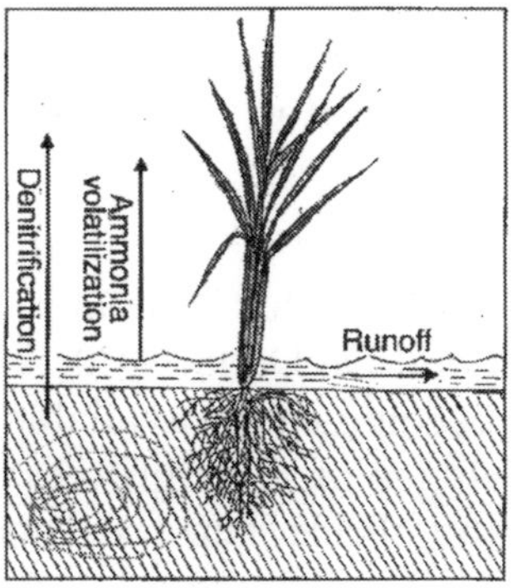

Fig. 32. Environmentally sound deep-placed briquettes in soil

Broadcasting and / or incorporation of conventional fertilizers and / or improper or shallow placement of briquettes in soil are prone to serious losses and, therefore, their fertilizer use efficiency is only up to 40 percent in farmers fields (Savant and De Datta, 1982). The losses of fertilizers cause environmental pollution (Fig. 33).

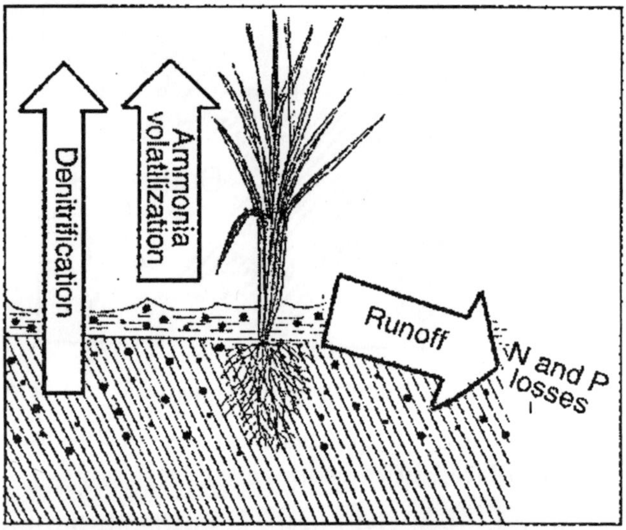

Fig. 33. Environmental pollution due to broadcasting and /or surface application of fertilizers

Briquettes are not effective if they are placed at less than 5 cm soil depth (Fig. 34). Under these conditions, losses of N due to ammonia volatilization and N and P due to surface run off are enhanced (Savant and De Datta, 1982; Savant *et al.*, 1982). Thereby, fertilizer pollution is increased with decrease in agronomic efficiency.

Briquettes may not be effective if they are placed at more than 10 cm soil depth (Fig. 35). This is because it requires too long a time for N (Savant and De Datta, 1980; Savant *et al.*, 1982) and P in briquettes to become easily available to the rice seedlings during the early stages of their growth. This may adversely affect the initial stand of rice crop and, therefore, the final yield.

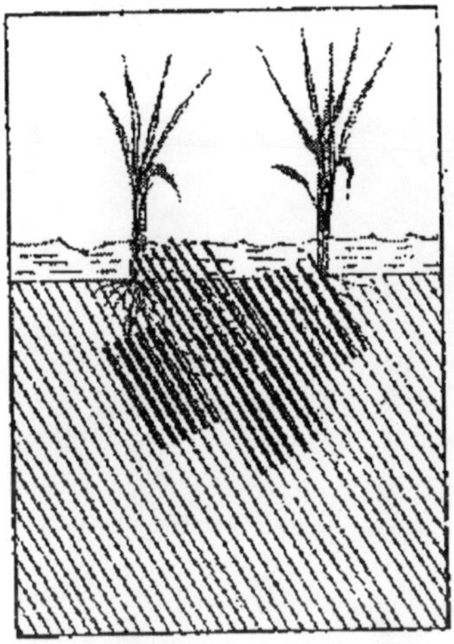

Fig. 34. Briquettes placed less than 5 cm soil depth : not efficient placement

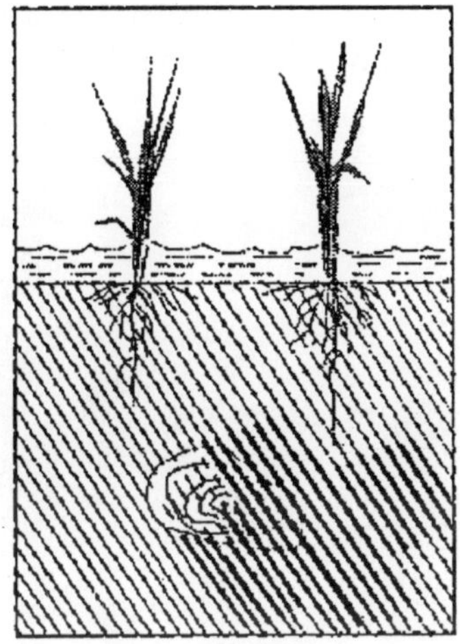

Fig. 35. Briquettes placed more than 10 cm soil depth : not efficient placement

- **Deep-placed Urea-DAP briquettes can eliminate fertilizer losses and pollution**

Placing the briquettes up to 7 to 10 cm depth in the soil reduces fertilizer losses up to 5 percent and as a result, its fertilizer use efficiency is significantly increased up to 70-80 percent.

- **Deep-placed briquettes do not promote weed growth**

Due to deep placement of N and P in the briquettes at 7 to 10 cm soil depth, weed growth is not promoted (Fig. 36). Therefore, there is less weed problem (Daftardar *et al.*, 1997 a).

Broadcasting or incorporation of fertilizers remain in floodwater and on surface soil. They stimulate weed growth during early growth of rice seedlings after transplanting and, therefore, there is a serious weed problem (Fig. 37).

- **Briquettes can be deep-placed in soil with standing floodwater**

Briquettes can be inserted at the rate of one briquette for every four hills in controlled transplanted rice field having 4 to 6 cm floodwater (Photo 12) and the same higher fertilizer use efficiency is achieved (Bulbule *et al.*, 1995; Savant *et al.*, 2000).

Photo 12. A worker inserting briquettes in the soil with standing floodwater

8.5 Rice soils generally with medium to fine texture are suitable for briquette placement

Rice soils with medium to fine texture (cation exchange capacity greater than 10 c mol (p^{+}) kg^{-1} generally have small to negligible leaching losses of fertilizer N. Such soils are suitable for deep placement of Urea-DAP briquettes (Fig. 38) (Misra *et al.*, 1995 ; Savant and Stangel, 1990).

Soils with coarse or sandy texture (cation exchange capacity lower than 10 c mol (p^{+}) kg^{-1}) have high leaching losses of fertilizer N. Such soils for example Punjab soils, are not suitable for deep placement of Urea-DAP briquettes (Fig. 39) (Savant and Stangel, 1990).

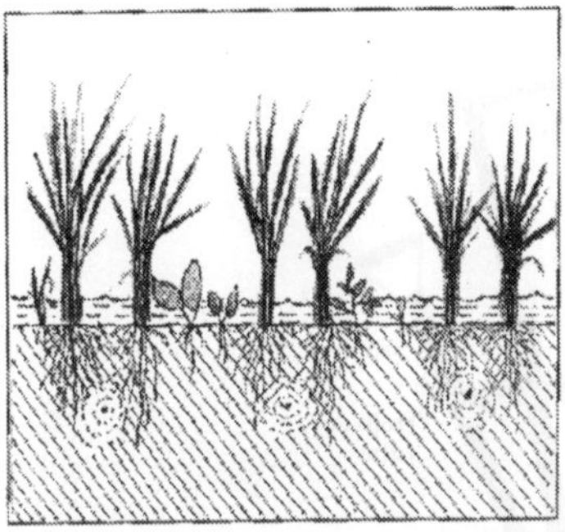

Fig. 36. Deep-placed (7-10 cm) briquettes in soil : less weeds

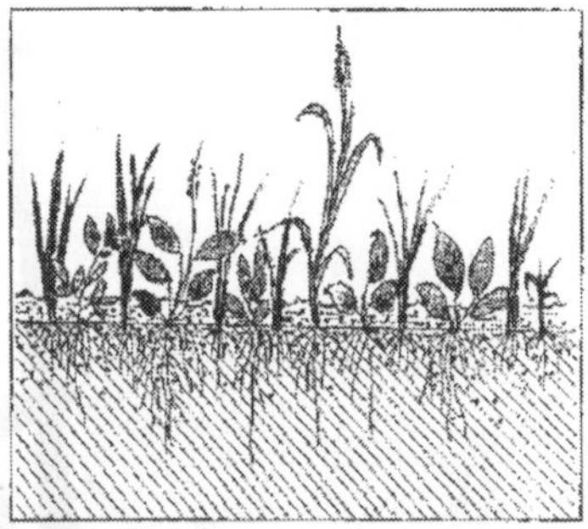

Fig. 37. Broadcast or incorporation of fertilizers in soil : more weeds

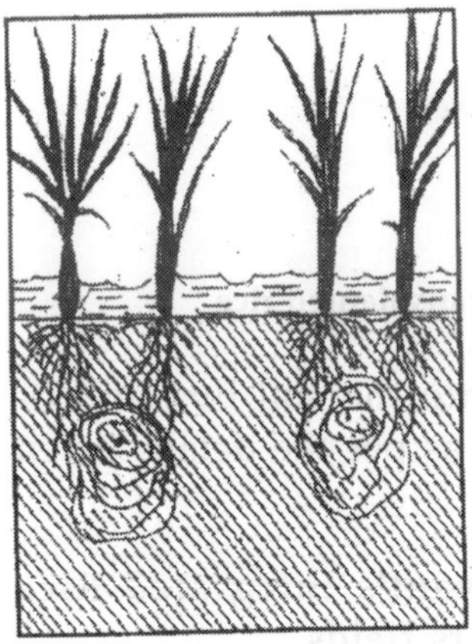

Fig. 38. Suitable soil conditions for briquette application

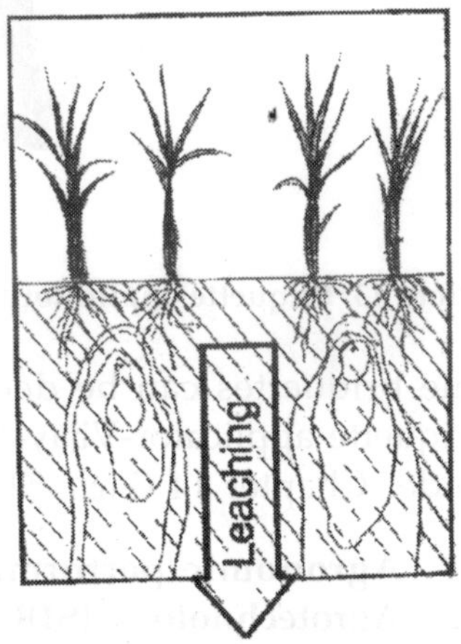

Fig. 39. Unsuitable soil conditions for briquette application

8.6 Equal opportunity for training to male and female workers are essential

The four practices of SIRA are simple and user friendly. Some of them require relatively less energy or strength under field conditions and, therefore, are suitable for women workers.

The field experiences have shown that female farmers perform field jobs such as transplanting, briquette placement etc. more efficiently than male workers. Therefore, male and female workers should be given equal importance in training and field practices (Photo 13) (Savant, 2001 a).

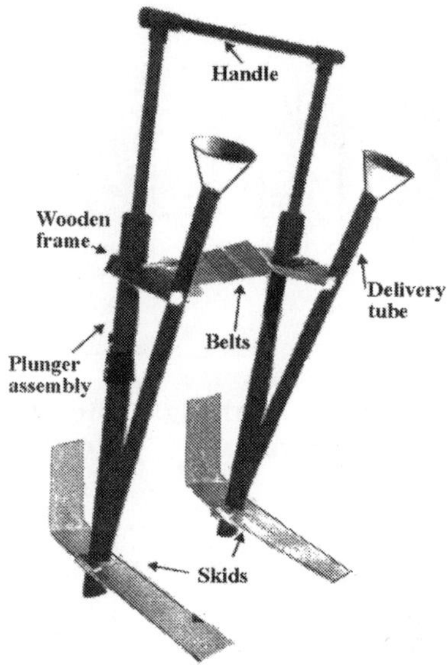

Photo 13. Briquette applicator (Vijay-1) for transplanted rice

The briquettes can be deep-placed into the soil by hand or briquette applicator-Vijay-1 (Photo 13) as per the details given under Chapter 14 (Figs. 51 and 54).

8.7 Agronomic performance of Savant's Integrated Rice Agrotechnology (SIRA) on farmers fields

8.7.1 Non Government Organisation (NGO)

8.7.1.1 Janaseva Foundation (Silvassa), Pune, *kharif* 2005-2007 (trials conducted in Gujarat State)

Using holistic approach, Savant's Integrated Rice Agrotechnology (SIRA) was developed (see Fig. 15). This SIRA has potential to increase present rice production with less fertilizers, therefore, less environmental pollution but more yield (Daftardar *et al.*, 1997 a ; Savànt, 2001 a). The SIRA has four main steps. (1) Recycling of rice plant residues, (2) Limited *gliricidia* green manuring using agroforestry approach, (3) Controlled transplanting of rice seedlings in both horizontal directions (at 25 hills m^{-2}) and (4) Deep placement of Urea-DAP briquettes in transplanted rice (one briquette for every four hills). The SIRA has been tested and found to increase rice yields of small, especially tribal farmers of Maharashtra for more than 10+ years (Savant, 2002; Savant and Daftardar, 2000; Savant, 2005). The results of farmer managed on-farm trials of SIRA in tribal areas of Surat, Valsad and Dang districts of Gujarat State during 2005, 2006 and 2007 *kharif* seasons are as follows (Anonymous, 2007a).

Kharif 2005 to 2007 : (Farmers' method)

The participating farmers carried out transplanting of rice seedlings either in line or in random manner. Average number of hills m^{-2} was 34. The dose and time of application of fertilizer varied from farmer to farmer. They applied either equal, more or less fertilizers as compared to the dose of 120 kg N (in two splits, approx. 60 kg N ha^{-1} at transplanting and 60 kg N ha^{-1} at 1 to 2 months after tillering stage) and 30 kg P_2O_5 ha^{-1} recommended by State Department of Agriculture, Gujarat (Anonymous, 2007a).

Kharif 2005 : SIRA method

The SIRA was explained to the tribal rice farmers in four tahsils viz. Songadh, Vyara, Uchhal and Mandvi by conducting group meetings in villages and with the help of hand-outs and booklets. On-farm trials were conducted on the fields of 545 farmers who volunteered to participate from 47 villages covering all the four tahsils. Agroclimatically, the trials were conducted on Western Ghat Zone or Sub-Montane Zone of Sahyadri and soils were mostly Entisols and sometimes tending towards Inceptisols. Average annual rainfall varied from 1600 to 2000 mm. The participating farmers were given training and then they grew rice by SIRA method in 100 m^{-2} (0.01 ha) area. They grew rice by traditional Farmers' method in the remaining area in their rice plots (to serve as control treatment). All the farmers used the same hybrid or improved variety of rice for the two treatments (Anonymous, 2007a).

SIRA method : Out of four agropractices, only two agropractices namely 3 and 4 of SIRA were used by the farmers. Controlled transplanting was done at .....25.....15.....25......15.... cm spacing in both directions with 25 hills m^{-2} (Step 3). One 2.7 g Urea-DAP briquette (Savant and Stangel, 1998) was deep-placed (7 to 10 cm) at the centre of each square (15 x 15 cm) of hills. The fertilizer dose was 58 kg N and 29 kg P_2O_5 ha^{-1} (Step 4). Agropractice 1 (recycling rice crop residues) and agropractice 2 (limited *gliricidia* leaves - about 3 t ha^{-1} incorporation in the soil during final puddling as green manure) of SIRA could not be adopted.

***Kharif* 2006 : SIRA method**

On-Farm SIRA Trials : On-farm trials were conducted voluntarily by 509 farmers in 48 villages in 4 tahsils of Surat district viz. Songadh, Vyara, Uchhal and Valod and by 502 farmers in 33 villages in 3 tahsils of Valsad district viz. Dharampur, Kaprada and Pardi during *kharif* 2006. The agroclimate and the same treatments given above for *kharif* 2005 were followed for the *kharif* 2006 season except for the following two points (Anonymous, 2007a).

1) In *kharif* 2006, rice hull black ash (about 1 kg ash m^{-2} in effective rice bed only) was used to supply silicon, an useful element, for the healthy growth of rice seedlings (Agropractice 1 of SIRA).
2) In *kharif* 2006, SIRA method was used on 500 m^{-2} (0.05 ha) while the remaining area (0.05 ha or more) in the plot was used for Farmers' method.

***Kharif* 2007 : SIRA method**

On-farm trials were conducted voluntarily by 176 farmers in villages in 4 tahsils of Dang district and by 45 farmers in villages of 3 tahsils of Valsad district viz. Dharampur, Kaprada and Pardi during *kharif* 2006. The agroclimate and the same treatments given above for *kharif* 2005 were followed for the *kharif* 2007 season. All the rice cultivation practices were similar for Farmers' method and SIRA method (Anonymous, 2007a).

Crop yields : It was possible to record rice yields of 209 farms in Surat district in 2005, 281 farms in Surat district and 217 farms in Valsad district in 2006. The rice yields increased by 42 percent in Surat district in 2005 (Table 2) and by 36 percent and 39 percent in Surat and Valsad districts, respectively in 2006 (Table 3) and 49 percent in Dang district while 36 percent in Valsad district in 2007 (Table 4) (Savant, 2002; Anonymous, 2007 a). When productivity by Farmers' method classified and yield increase due to SIRA for each class considered, it was found that yield increase was more for lower productivity class than for higher productivity class. This shows that farmers and farms with lower yield levels are benefitted more by using SIRA. When value of additional produce due to SIRA and extra expenses in cost of cultivation due to SIRA (value/ cost ratio) were considered, the farmers with lower yield levels obtained higher returns than the farmers with higher yield levels (Table 2). This yield and economic assurance to poor productivity farms and farmers are a characteristic feature of SIRA which this study has clearly revealed.

A decade of experience of SIRA has shown that it always

Table 2. Rice yields of on-farm trials of SIRA on farmers' fields in Surat district (Gujarat), *kharif* 2005

Productivity class (q ha^{-1})	No. of farmers who recorded yields	Mean rice yield (q ha^{-1})			Value (Rs. ha^{-1}) *		Returns per rupee spent, value/ cost ratio **
		Farmer's method	SIRA	% increase	Farmer's method	SIRA	
31 to 40.9	20	36.3 (2.68)	61.2 (2.69)	69	19965	33660	6.5
41 to 50.9	100	45.2 (2.29)	66.5 (3.09)	47	24860	36575	5.6
51 to 60.9	73	54.9 (3.96)	74.9 (3.96)	36	30195	41195	5.3
61 to 70.9	16	68.3 (5.27)	82.3 (3.72)	20	37565	45265	3.7
Total 209 (out of 545)		**49.6 (8.17)**	**70.2 (6.73)**	**42**	**27115**	**38610**	**5.5**

* Rate of rice : Rs. 550 per q.

** Returns or Recovery = value / cost ratio.

Figures in bracket indicate standard deviation.

Table 3. Rice yields of on-farm SIRA trials on farmers' fields in Surat and Valsad districts (Gujarat), *kharif* 2006

Productivity (q ha^{-1}) Class	Surat district				Valsad district			
	No. of farmers who recorded yields	Mean rice yield (q ha^{-1})			No. of farmers who recorded yields	Mean rice yield (q ha^{-1})		
		Farmers' method	SIRA	% increase		Farmers' method	SIRA	% increase
21.0 to 30.9	59	23.9 (4.72)	36.4 (10.90)	52.3	36	27.3 (2.76)	41.0 (5.2)	50.2
31.0 to 40.9	75	37.8 (2.84)	59.1 (8.12)	56.3	125	35.6 (3.03)	49.9 (5.48)	40.2
41.0 to 50.9	87	45.9 (4.49)	64.0 (6.59)	39.4	53	43.7 (2.68)	58.9 (5.14)	34.8
51.0 to 60.9	47	54.4 (2.67)	67.8 (4.33)	24.6	3	54.5 (4.65)	69.7 (2.75)	27.9
61.0 to 70.9	13	65.3 (3.11)	75.7 (4.79)	15.9	-	-	-	-
	Total 281 (out of 508)	42.7 (11.89)	58.1 (14.49)	36.1	Total 217 (out of 502)	36.5 (6.30)	50.9 (8.44)	39.5

Figures in bracket indicate standard deviation.

Table 4. Rice yields of on-farm SIRA trials on farmers' fields in Dang and Valsad districts (Gujarat), *kharif* 2007

Productivity (q ha^{-1}) Class	Dang district				Valsad district			
	No. of farmers who recorded yields	Mean rice yield (q ha^{-1})			No. of farmers who recorded yields	Mean rice yield (q ha^{-1})		
		Farmer's method	SIRA	% increase		Farmer's method	SIRA	% increase
21.0 to 30.9	56	26.71 (1.98)	38.21 (4.86)	43	6	25.33 (3.27)	38.00 (1.20)	50
31.0 to 40.9	63	36.51 (2.79)	56.49 (8.50)	54	19	35.79 (3.64)	50.95 (5.30)	42
41.0 to 50.9	49	44.04 (2.36)	66.20 (7.31)	50	15	46.40 (2.02)	60.67 (4.82)	30
51.0 to 60.9	8	56.50 (3.38)	78.25 (1.01)	38	5	52.80 (1.79)	66.40 (6.63)	25
Total 176		36.40 (8.43)	54.37 (14.22)	49.0	Total 45	39.82 (8.80)	54.18 (9.83)	36

produces higher yield than the control (full recommended dose of fertilizers) (Savant, 2002). It is also almost always observed that the yield increase due to SIRA is substantial i.e. not less than 1 t ha^{-1}. In the present study, the per hectare average yield increased ranging from 10.2 qha^{-1} to 24.9 qha^{-1} over and above the yield obtained by the Farmers' method. This confirms the above claim (Savant, 2001 a). In terms of income, the minimum yield increase of 1 t ha^{-1} due to SIRA assures the farmer of more than Rs. 5000/- ha^{-1} profit over Farmers' method at current market rate of rice.

Kharif 2008 :

On-farm trials were conducted on lowland rice during the *kharif* 2008 on the farmers' fields. The trials were conducted on the fields of 137 farmers in Vasada tahsil of Navasari district, 172 farmers in Ahawa tahsil of Dang district and 218 farmers in Dharampur tahsil of Valsad district (Gujarat State) by Janaseva Foundation (Silvassa), Pune (Anonymous, 2009 a). The package of practices followed by the farmers under Farmers' method and SIRA method were the same as described above under section 8.7.1.1.

Crop yields :

In Navasari district 137 farmers from Vasada tahsil conducted lowland rice trials during the *kharif* 2008. They used I.R. 28, Jaya and Masuri rice cultivars. Out of 137 farmers, 101 farmers recorded rice yields under productivity class of 31- 40.9 qha^{-1} (mean yields of 35.5 qha^{-1} and 55.8 qha^{-1}, respectively under Farmers' method and SIRA method increasing the yields by about 57 %) (Table 5). The calculated 't' value was highly significant (22.18**) indicating that SIRA method was superior to Farmers' method (Anonymous, 2009 a).

In Dang district, 172 farmers from Ahawa tahsil conducted rice trials with I.R.28, Jaya, Masuri, Guj.7, 6201, and 6441 cultivars. Out of 172 farmers, 63 farmers recorded rice yields under productivity class of 31- 40.9 qha^{-1} (mean yield 36.8 qha^{-1} under Farmers' method and 61.1 qha^{-1} under SIRA method with an increase of 66%). This was followed by 57 farmers under 51-60.9 qha^{-1} and 47 farmers under 41-50.9 qha^{-1} productivity class increasing the yields by about 83% and 70%, respectively. The over all increase in rice yield in SIRA method was 53.7%. The calculated 't' value was highly significant indicating the superiority of SIRA method over the Farmers' method (Table 6).

In Valsad district, 218 farmers conducted rice trials using I.R. 28, Jaya and Masuri cultivars. Out of 218 farmers, the yields of 138 farmers were under 31- 40.9 qha^{-1} productivity class with a mean yields of 36.1 qha^{-1} and 46.0 qha^{-1} under Farmers' and SIRA methods, respectively (Table 7). The over all increase in yield in SIRA method was 27.6%. The calculated 't' value was highly significant indicating the superiority of SIRA method.

In all 528 farmers from Navasari, Dang and Valsad districts conducted rice trials to compare Farmers' method and SIRA method. The mean rice yields were 39.6 qha^{-1} and 56.9 qha^{-1}, respectively under Farmers' and SIRA methods (Table 8). Calculated 't' value was highly significant indicating the superiority of SIRA method. This increase in rice yields were attributed to increase in number of grains per panicles and grain weight (Anonymous, 2009 a).

The SIRA method is thus useful for increasing the rice yields of the farmers. The increase in yields were higher in the case of farmers having lower yields.

8.7.1.2 Trials conducted by Krishi Vigyan Kendras (KVKs): Three KVKs conducted on-farm trials during *kharif* 2006 and reported increase in rice yield by 25 percent (KVK, Waghai, dist. Dang), 42 percent (KVK, Ambheti, dist. Valsad) and 23 percent (KVK, Vyara, dist. Surat) due to SIRA over full dose of 120 kg N ha^{-1}. The former two KVKs trials reported uniform crop stand and dark green foliage and better tillering in SIRA as compared to the control. The trial at Vyara was affected by cyclone but, at all the KVKs, SIRA with 58 kg N ha^{-1} recorded higher yield over the university recommended dose of 120 kg N ha^{-1} (Anonymous, 2007a).

- **Other benefits :** The participating farmers experienced and reported saving of seed and seedlings, nitrogen fertilizer and labour for planting and harvesting in SIRA method. They also reported less weeds in SIRA plots due to deep-placed fertilizer which resulted in saving of labour for weeding.

Table 5. Rice yields of on farmers' fields in Vasada tahsil of Navasari district (Gujarat), *kharif* 2008

Productivity class (q ha^{-1})	No. of farmers	Navasari district Mean rice yield (qha^{-1})			
		Farmers' method	SIRA	% increase	' t 'cal.
21 to 30.9	14	26.8 (2.68)	40.1 (8.12)	49.5	22.18**
31 to 40.9	101	35.5 (2.89)	55.8 (7.02)	57.0	
41 to 50.9	21	44.9 (2.50)	65.5 (7.64)	45.4	
51 to 60.9	1	52.0 (5.52)	68.0 (0.0)	30.0	
Total	137	36.1 (5.52)	55.8 (8.79)	54.5	

Figures in bracket indicate standard deviation.
Rice cultivars used : I.R. 28, Jaya, Masuri.

Table 6. Rice yields of on farmers' fields in Ahawa tahsil of Dang District (Gujarat), *kharif* 2008

Productivity class (q ha^{-1})	No. of farmers	Dang district Mean rice yield (qha^{-1})			
		Farmers' method	SIRA	% increase	' t 'cal.
21 to 30.9	4	28.0 (0.0)	46.0 (7.66)	64.2	22.04**
31 to 40.9	63	36.8 (3.03)	61.1 (7.25)	66.0	
41 to 50.9	47	45.9 (2.02)	70.4 (6.80)	53.3	
51 to 60.9	57	55.5 (3.12)	80.0 (6.38)	45.0	
61 to 70.9	1	64.0 (0.0)	80.0 (0.0)	25.0	
Total	172	45.4 (8.74)	69. 8 (11. 57)	53.7	

Figures in bracket indicate standard deviation.
Rice cultivars used : I. R. 28, Jaya, Masuri, Guj. 7, 6201, 6444.

Table 7. Rice yields on farmers' fields in Dharampur tahsil of Valsad District (Gujarat), *kharif* 2008

Productivity class (q ha^{-1})	No. of farmers	Valsad district			
		Mean rice yield (qha^{-1})			
		Farmers' method	SIRA	% increase	't 'cal.
21 to 30.9	32	27.0 (2.03)	37.4 (3.31)	38.5	13.72**
31 to 40.9	138	36.1 (3.07)	46.0 (5.47)	27.4	
41 to 50.9	37	45.6 (1.99)	57.1 (7.64)	25.2	
51 to 60.9	11	53.8 (2.06)	65.1 (5.39)	216	
Total	**218**	**37.2 (6.91)**	**47.5 (8.65)**	**27.6**	

Figures in bracket indicate standard deviation.
Rice cultivars used : I. R. 28, Jaya, Masuri.

Table 8. Rice yields on farmers' fields in Navasari, Dang and Valsad districts (Gujarat), *kharif* 2008

District	No. of farmers	Rice yield (qha^{-1})			
		Farmers' method	SIRA	% increase	't 'cal.
Navasari	137	36.1 (5.52)	55.8 (8.79)	53.6	22.04**
Dang	172	45.4 (8.74)	69.8 (11.56)	51.5	
Valsad	218	37.2 (6.91)	47.5 (8.65)	27.6	
Total	**528**	**39.6 (8.31)**	**56.9 (13.64)**	**43.7**	

Figures in bracket indicate standard deviation
Rice cultivars used : I.R. 28, Jaya, Masuri, Guj.7, 6201, 6444

8.7.1.3 Janaseva Foundation (Silvassa), Pune, *kharif* 1993 - 2001 (trials conducted in Maharashtra State)

Highlights :

- In view of the utility and suitability of Savant's Integrated Rice Agrotechnology (SIRA) for agricultural extension in tribal areas, the Janaseva Foundation (Silvassa), Pune for the first time, conducted adaptive trials on tribal farmers' fields during *kharif* seasons of 1993 to 1995 (Table 9) (Savant, 2002 ; Savant and Daftardar, 2000; Anonymous, 2001 a).

- Grain and straw yields of rice were substantially increased by controlled transplanting and using Urea-DAP briquettes (Practices 3 and 4) irrespective of variations in yearly rainfall, varieties of paddy and soils used. These yields of rice grain and straw were higher than the yields obtained by Farmers' method and local method with chemical fertilizers.
- For every rupee involved for the Urea-DAP briquette use, the tribal farmers received in returns additional yields of rice grain and straw worth Rs. 5 to 10 in the same season.

Table 9. Adaptive rice trials using Urea-DAP briquettes on tribal farmers' fields, *kharif* 1993 to 1995

<table>
<tr><th rowspan="2">Year</th><th rowspan="2">No. of trials</th><th rowspan="2">Crop management</th><th colspan="2">Rice yield (q ha^{-1})</th><th rowspan="2">Districts</th></tr>
<tr><th>Grain</th><th>Straw</th></tr>
<tr><td rowspan="3">1993</td><td rowspan="3">26</td><td>Farmers' method</td><td>24</td><td>41</td><td rowspan="3">Thane, Raigad, Nasik and Pune</td></tr>
<tr><td>Local transplanting and use of urea-superphosphate</td><td>30</td><td>45</td></tr>
<tr><td>Controlled transplanting and use of Urea-DAP briquettes (Practices 3 and 4)</td><td>46</td><td>67</td></tr>
<tr><td rowspan="3">1994</td><td rowspan="3">51</td><td>Farmers' method</td><td>27</td><td>42</td><td rowspan="3">Thane, Raigad, Dadara, Nagar Haveli, Valsad, Nasik, Pune and Ahmednagar</td></tr>
<tr><td>Local transplanting and use of urea-superphosphate</td><td>31</td><td>44</td></tr>
<tr><td>Controlled transplanting and use of Urea-DAP briquettes (Practices 3 and 4)</td><td>39</td><td>53</td></tr>
<tr><td rowspan="3">1995</td><td rowspan="3">40</td><td>Farmers' method</td><td>27</td><td>32</td><td rowspan="3">Thane, Raigad, Nasik, Pune and Valsad</td></tr>
<tr><td>Local transplanting and use of urea-superphosphate</td><td>30</td><td>32</td></tr>
<tr><td>Controlled transplanting and use of Urea-DAP briquettes (Practices 3 and 4)</td><td>41</td><td>46</td></tr>
</table>

- In agrotechnology transfer trials on farmers' fields during 1996 to 2001, use of controlled transplanting and Urea-DAP briquettes (Practices 3 and 4) increased grain yields by 35 to 83 percent (Table 10) (Savant, 2002 ; Savant and Daftardar, 2000 ; Anonymous, 2001 a).
- Several NGOs in Maharashtra State conducted several trials on farmers' fields by using controlled transplanting and Urea-DAP briquettes (Practices 3 and 4). The farmers are now convinced about efficiency of SIRA in increasing yields of rice grain and straw (Table 11). This has resulted in widespread propaganda and technology transfer of SIRA by NGOs (Savant, 2002; Savant and Daftardar, 2000 ; Anonymous, 2001 a).

Table 10. Technology transfer rice trials using Savant's Integrated Rice Agrotechnology (Practices 3 and 4) on tribal farmers' fields, 1996 to 2001

Year	No. of trials	Grain yield (q ha $^{-1}$)			Districts
		Use of Urea-DAP briquette	Farmers' method	Increase in yield (%)	
1996	68	57	34	68	Raigad, Thane, Pune, Dhule, Ahmednagar, Dadara and Nagar Haveli
1997	36	55	39	41	Raigad, Thane and Pune
1998	112	50	37	35	Nasik
1999	52	46	29	59	Thane, Nandurbar and Pune
2000	196	37	23	61	Thane, Nandurbar and Pune
2001	39	58	33	75	Thane
	100	57	40	42	Nandurbar
	3	68	37	83	Dhule

Table 11. Technology transfer rice trials on farmers' fields using Urea-DAP briquettes after controlled transplanting, 2000-2001

NGO	Details of trials				Average yield (q ha^{-1})				Increase in yield (%)
					Use of Urea-DAP briquettes		Farmers' method		
	Place	Period	No. of farmers	Paddy variety	Grain	Straw	Grain	Straw	
Dist. Ratnagiri									
Care India, Kokan Integrated Development Project, At. Abloli, Tah. Guhagar	Abloli	2000 and 2001	320+	Jaya, Suvarna, PHB and Sahyadri	60-120	-	25-30	-	100-300
Pramod Mandavkar At. Jalgaon, Tah. Dapoli, Dist. Ratnagiri	Jalgaon Tah. Dapoli	2001	15	Sahyadri, Suvarna, Proagro, Karjat-3 and Ratna	63	94	45	60	-
Prashant Shirgaonkar Z.P. member, At Palshet, Tah. Guhagar, Dist. Ratnagiri	Palshet, Tah. Guhagar,	2001	44	Sahyadri, Masuri, Karjat-3, Karjat-4 and Proagro	90	300	40	250	125
Other States									
Suyash Charitable Trust, Pirangut, Tah. Mulshi, Dist. Pune	Chhattisgad Orissa	2001	38	I R 36, Kalinga and Mahamaya	63	-	31	-	100

8.8 State Department of Agriculture / Zilla Parishad (Maharashtra)

8.8.1 Superintending Agriculture Officer, Pune and Ahmednagar districts

Highlights :

- Field demonstrations of SIRA (each on 100 to 300 m^2 area) were organised under Integrated Cereal Crop Development Programme in Pune and Ahmednagar districts during the *kharif* 2001.
- Controlled transplanting (25 hills m^{-2}) (Practice 3) and efficient use of Urea-DAP briquettes (Practice 4) were adopted.
- In Pune district, rice grain and straw yields increased by 21 to 85 percent and 16 to 128 percent, respectively (Table 12) (Savant, 2002; Anonymous, 2001 b).
- In Ahmednagar district, increases in rice grain and straw yields were 22 percent and 48 percent, respectively (Table 12) (Anonymous, 2001c).

Reactions of Farmers :

- Use of Urea-DAP briquettes enhanced growth and increased number of tillers per hill. Rice yield increased with less seed rate and less fertilizers.
- Integrated use of improved variety or hybrid rice with Urea-DAP briquettes would make rice farming profitable and would improve economic conditions of farmers.
- Farmers appreciated the results of the briquette technology and decided to adopt SIRA on a large scale from next year.

Table 12. Results of rice trials by using Urea-DAP briquettes, *kharif* 2001

District / Tahsil	No. of trials	Variety	Mean rice yields (q ha $^{-1}$)				Increase in yield (%)	
			Traditional method		Use of Urea-DAP briquettes			
			Grain	Straw	Grain	Straw	Grain	Straw
Pune district								
Bhor	12	Pusa Basmati	22	32	27	37	23	16
Mulsi	12	Pusa Basmati	25	35	36	47	44	34
Haveli	9	Indrayani	28	29	34	66	21	128
Khed	11	Indrayani	21	52	35	63	67	21
Maval	12	Sahyadri	39	83	72	128	85	54
Junnar	14	Indrayani	19	43	23	60	21	40
Ambegaon	11	Sahyadri	17	44	22	66	29	50
Velha	12	Sahyadri	31	40	54	57	74	43
Ahmednagar district								
Akole	155	Different varieties	27	29	33	43	22	48

8.8.2 Tahsil Agriculture Officer (Ambegaon) Junnar, Dist. Pune, *kharif* 2000

Highlights :

- Forty seven trials were conducted on farmers' fields during *kharif* 2000 season. Yield data of 8 trials are reported in Table 13 (Savant, 2002; Anonymous, 2001 d).

8.8.3 Tahsil Agriculture Officer, Velha, Dist. Pune

Kadave is a small village having a population of 1200 people in Velha tahsil of Pune district. It is situated in a hilly remote place. The soils of the village are laterite with an average rainfall of 2000 mm. Rice is the main crop of the village. Three years ago i.e. in 2005, regional staff of the Agriculture Department organised a meeting of the villagers (farmers) and gave information about the Savant's Integrated (*Charsutri*) Rice Agrotechnology. The farmers were convinced and adopted this

Table 13. Results of rice trials in Pune district, *kharif* 2000

Name of village	Name of farmer	Rice cultivar	Rice grain yield, q ha^{-1}	
			Traditional method	Use of Urea-DAP briquette
Phulvade	Gangaram Limaji Hile	Sahyadri	47	95
Mapoli	Rama Krishna Lohkare	Sahyadri	32	65
Mapoli	Bhimji Dongru Lohkare	Sahyadri	39	79
Mapoli	Vithal Valhu Pote	Sahyadri	31	57
Supekar	Dyanu Bhokar Rabhe	Sahyadri	32	71
Supekar	Kisan Rama Bhagat	Sahyadri	40	74
Supekar	Chandrakant Kisan Kolte	Indrayani	42	126
Ghodegaon	Sakharam Mhatarba Darekar	Sahyadri	53	114
		Average	40	85

agrotechnology using Indrayani rice variety. In the first year, the average rice yields of the farmers increased from 2.2 t ha^{-1} to 4.5 t ha^{-1}. The staff of the State Agriculture Department, Pune region provided complete guidance to the farmers (Agro One, 2008 a). In the second year i.e. in 2006, the average rice yields of the farmers increased to 6.2 t ha^{-1}, while in the third year (in 2007) it increased to 7.5 t ha^{-1}. About 70% farmers are adopting Savant's Integrated (*Charsutri*) Rice Agrotechnology. Due to this progress in agriculture the village has been recently given "Krishigram" award by the State Department of Agriculture. The farmers adopted co-operative seedling nursery on raised seedbeds for which State Department of Agriculture provided seeds, insecticides and subsidy for vermicompost. Because of co-operative nursery there was a saving of time, labour and money (Photos 14 and 15).

Photo 14. Participants in integrated method of rice cultivation at Kadave are (from left) Shri. Govind Parjane, Asstt. Agriculture Officer; Shri K.V. Deshmukh, Joint Director of Agriculture, Shri. Sanjay Hiwale, Mandal Agriculture Officer, Pune region

Photo 15. Visit of Agriculture Officers to seedling nursery of the farmer, Shri Datta Marathe at Kadave, distt. Pune

8.8.4 Tahsil Agriculture Officer, Mulshi, Dist. Pune

Co-operative Seed Production Programme Through *Charsutri* (SIRA) Rice Cultivation, *kharif* 2008 :

Nande is a small village in Mulshi tahsil of Pune district where lowland rice is the main crop grown. Farmers from Nande and adjoining villages grow rice as per their traditional practice. However, there are many farmers who follow *Charsutri* (SIRA) method of rice cultivation and get good rice yields. Bhairavnath Krishi Vidnyan Mandal , Nande have made available thousands of quintals of good quality rice seeds to the farmers during the past 3-4 years. Considering the positive approach, Bhairavnath Krishi Vidnyan Mandal was given registration and permission for sale of seeds.

During the *kharif* 2008, Mulshi Tahsil Agriculture Officer planned to implement co-operative rice seed production under Maharashtra Increased Agricultural Productivity Programme in Nande village through Bhairavnath Krishi Vidnyan Mandal. The objective of this programme was to get pure and good quality of the rice seeds to the farmers of that area at reduced price. Thus, co-operative rice seedling nursery (cv. Indrayani) was established to reduce the cost of cultivation and to get good, healthy seedlings. This programme was implemented on the fields of 24 farmers. All these 24 fields (plots) were registered for Indrayani rice seed production programme with Seed production Officer of the Agriculture Department. Mahatma Phule Krishi Vidyapeeth, Rahuri provided Breeder's Indrayani rice seeds (Agro One, 2009). Many farmers, Officers of the Agriculture Department and MPKV Rahuri visited the seed production plots (Photo 16).

Thus, during the *kharif* 2008, 24 farmers produced 71 tons of rice seeds (Indrayani variety) through Bhairavnath Krishi Vdnyan Mandal under the guidance of Agriculture Department. Out of 71 tons of seeds, 50 tons were made available to other farmers of Nande and adjoining villages on sale. In this programme the cost of production was Rs. 40000/- ha^{-1} and receipts of Rs. 102000/- ha^{-1} with a profit of Rs. 62000/- ha^{-1}.

Photo 16. ***Charsutri*** **rice (cv. Indrayani) crop ready for harvest at Nande village, Mulshi tahsil, Pune district,** ***kharif*** **2008**

8.8.5 Superintending Agriculture Officer, Kolhapur district, *kharif* 2001

Highlights :

- In Kolhapur district transplanted rice is a practice in tahsils on western side of Kolhapur (Savant, 2002; Anonymous, 2001e). Ratnagiri 24, Ratnagiri 500, Jaya, Karjat 1, Radhanagari185-2 and Sahyadri rice cultivars are used.
- In Kolhapur district, 2237 beneficiary farmers used Urea-DAP briquettes in trial plots on an area of 516 ha. Transplanting by farmers' method along with fertilizer application resulted in grain and straw yields of 26 q ha^{-1} and 66 q ha^{-1}, respectively. Use of Urea-DAP briquettes however, recorded grain and straw yields of 34 and 85 q ha^{-1} which were higher than those obtained by Farmers' method by 31 and 29 percent, respectively (Savant, 2002 ; Anonymous, 2001e).

8.8.6 Superintending Agriculture Officer, Satara district, *kharif* 2001

- In Satara district, transplanted rice is grown in the tahsils of western sides. Pusa, Basmati 1, Basmati 370 and Sahyadri (hybrid) rice cultivars are grown (Anonymous, 2001 f).

- In Satara district, 3.94 tons of Urea-DAP briquettes were used in trial plots on an area of 230 ha and rice yields increased by 15 to 20 percent.
- Considering increase in response to use of Urea-DAP briquettes, fertilizer dealers / NGOs are actively participating in preparation and distribution of briquettes to farmers.

8.8.7 Sub-Divisional Agriculture Officer, Nasik, *kharif* 2001

Highlights :

- Cultivated area of rice in four tahsils of Nasik district viz., Igatpuri, Peth, Nasik and Sinnar were 16900, 8800, 4980 and 875 ha, respectively. The average production in these tahsils ranged from 13.6 qha^{-1} (in Nasik) to 18.1 qha^{-1} (in Igatpuri) (Savant, 2002; Anonymous, 2001 g).
- Forty one percent farmers used improved variety / hybrid rice, while 59% are still using local varieties.
- Controlled transplanting (Practice 3) followed by deep placement of Urea-DAP briquettes (Practice 4) increased rice yield appreciably (Table 14) (Savant, 2000; Anonymous, 2001 g).

Table 14. Effect of application of Urea-DAP briquettes on rice yields in Nasik district, 1999-2001

Year	Average productivity of tahsil 2000-2001	Use of Urea-DAP briquettes			
		Area (ha)	Quantity (t)	Average rice production ($q\ ha^{-1}$)	% increase
1999	13.6 to 18.2	282	26	40.3	100+
2000		309	53	25.5	39
2001		389	73	47.6	150+

- During *kharif* season of 2001, 533 farmers from 233 villages used 65 q of Urea-DAP briquettes in unaided farm trials on an area of 349 ha.

- During *kharif* season of 2002, 209 q of Urea-DAP briquettes are proposed to be used on an area of approximately 1629 ha (5.1 percent). It is proposed to have one Kranti Briquetter at Tahsil Seed Farm, Igatpuri and two in joint collaboration with Tribal Sale Purchase Society for smooth supply of briquettes for this programme.
- With a view of helping tribal farmers to adopt this technology, it is proposed to make available crop loans (50 percent inputs and 50 percent cash) through co-operative banks.

8.8.8 Sub-Divisional Agricultural Officer, Dapoli, Dist. Ratnagiri, *kharif* 2000

Highlights :

- Sixteen trials of SIRA were conducted on farmers' fields under Integrated Cereal Development Programme (Table 15) during 2000 *kharif* season (Savant, 2002; Anonymous, 2001 h).

Table 15. Results of trials on transplanted rice on farmers' fields, *kharif* 2000

<table>
<tr><th rowspan="2">Tahsil</th><th rowspan="2">Village</th><th rowspan="2">No. of trials</th><th rowspan="2">Variety</th><th colspan="2">Rice grain yield (q ha^{-1})</th></tr>
<tr><th>Use of Urea-DAP briquettes</th><th>Traditional method</th></tr>
<tr><td rowspan="3">Dapoli</td><td>Satere</td><td>4</td><td>Pro-Agro</td><td>59</td><td rowspan="3">30 to 55</td></tr>
<tr><td rowspan="2">Palgad</td><td>1</td><td>Karjat</td><td>54</td></tr>
<tr><td>1</td><td>Suvarna</td><td>48 to 52</td></tr>
<tr><td rowspan="3">Khed</td><td>Ambavali</td><td>3</td><td>Pro-Agro</td><td>48 to 52</td><td rowspan="3">20 to 30</td></tr>
<tr><td rowspan="2">Khavati</td><td>1</td><td>Pro-Agro</td><td>40</td></tr>
<tr><td>1</td><td>Palghar</td><td>37</td></tr>
<tr><td rowspan="2">Mandangad</td><td>Bhadevali</td><td>4</td><td>Palghar</td><td>42 to 45</td><td rowspan="2">25 to 30</td></tr>
<tr><td>Gote</td><td>1</td><td>Palghar</td><td>42 to 45</td></tr>
</table>

- Use of Urea-DAP briquettes increased rice yields.
- In Natunagar (Tah. Khed), farm trials were conducted on an area of 10 ha during summer season of 2000 (Savant, 2002; Anonymous, 2001 h).

8.8.9 Superintending Agriculture Officer, Ahmednagar District, *kharif* 2008

During the *kharif* 2008, 314 demonstration trials on rice were conducted on the farmers' fields in Akole tahsil of Ahmednagar district to compare the Farmers' method (traditional method) and *Charsutri* (SIRA) method of cultivation. The farmers used Indrayani, Poonam, Madhuvati, Vijaya, Rupali, Kolpi 248, Samruddhi, Sweta, Sonam Sahyadri, Ambemohar etc. rice cultivars (Anonymous, 2009 b). These trials were conducted to compare Farmers' method (traditional method) and *Charsutri* (SIRA) method of cultivation using Indrayani, Poonam, Madhuvati, Vijaya, Rupali, Kolpi 248, Samruddhi, Sweta, Sonam, Sahyadri, Ambemohar etc. rice cultivars. The results showed that rice yields of the farmers in lower productivity class (11- 20.9 qha^{-1}) increased from 16.4 qha-1in Farmers' method to 26.4 qha-1 in SIRA method increasing the yields by about 61% (Table 16). The over all increase in rice yields of the farmers was from 28.0 qha-1in Farmers' method to 28.0 qha-1 in SIRA method (about35 % increase). The calculated 't' value was highly significant indicating the superiority of SIRA method over the Farmers' method (Anonymous, 2009 b).

Table 16. Rice yields on farmers' fields in Akole tahsil of Ahmednagar district (Maharashtra), *kharif* 2008

Productivity class (qha^{-1})	No. of farmers	Mean rice yield (qha^{-1})			
		Farmers' method	SIRA	% increase	' t 'cal.
11 -20.9	166	16.4 (2.73)	26.4 (5.79)	53.6	26.74**
21 -30.9	148	25.6 (2.52)	29.9 (2.78)	51.5	
Total	314	20.8 (5.31)	28.0 (4.99)	27.6	

Figures in bracket indicate standard deviation.

Rice cultivars used : Indrayani, Poonam, Madhuvati, Vijaya, Rupali, Kolapi 248, Samruddhi, Sonam, Sweta, R-24, Shyadri, Ambemohar, Komal, Raibhog.

8.8.10 Agriculture Development Officer, Zilla Parishad, Kolhapur, *kharif* 2001

Highlights:

- In Kolhapur district, during *kharif* 2001 season, rice was transplanted on an area of 36000 ha.
- During 2000-01, a grant of Rs. 40,000/- was made available from Zilla Parishad cess fund for purchase of four Kranti Briquetters. This helped to supply 127 tons of Urea-DAP briquettes to farmers.
- One hundred twenty eight trials each on 300 m^2 area were conducted using improved rice varieties like Karjat-5, Jaya, R-24, Ratnagiri-711, Ratnagiri-1, IR-36, Indrayani and hybrid rice like MTU-1001, Sonam, Pro-Agro, Sahyadri etc. (Anonymous, 2001 i).
- In trial plots, average grain and straw yields of rice were 61.1 q ha^{-1} and 186.2 q ha^{-1}, respectively using controlled transplanting (Practice 3) and application of Urea-DAP briquettes (Practice 4). While the yields were 52.2 q ha^{-1} and 150.7 q ha^{-1}, respectively using farmers' methods. In short, even by using only Practices 3 and 4 of SIRA, the grain and straw yields of rice increased by 20 percent and 22 percent, respectively with less seeds, seedlings and fertilizers (Anonymous, 2001 i).

Reactions of NGOs :

There is a dire need to train NGOs, field workers and farmers (especially women) in controlled transplanting technique (Practice 3) and method of application of Urea-DAP briquettes (Practice 4).

8.8.11 Block Development Officer, Panchayat Samiti, Guhagar, Dist. Ratnagiri, *kharif* 2001

Highlights :

- During 2001 *kharif* season, 15 trials were conducted on farmers' fields using improved varieties like Masuri, Karjat-3, Karjat-4 and hybrid rice like Sahyadri.

- Average yields of rice grain and straw of improved varieties by farmers' method were 40-50 q ha^{-1} and 150-200 q ha^{-1}, respectively. Use of Urea-DAP briquettes however increased rice yields by 10 to 15 percent.
- Hybrid rice (Sahyadri) with Urea-DAP briquettes yielded 88 q ha^{-1} of grain (100 percent increase) and 265 q ha^{-1} of straw (Savant, 2002; Anonymous, 2001j).
- Integrated use of high yielding hybrid rice with SIRA would make rice farming a more profitable proposition.

8.9 Agricultural Universities in Maharashtra State

8.9.1 Dr. Balasaheb Sawant Konkan Krishi Vidyapeeth, Dapoli, Dist. Ratnagiri

- The SIRA developed by Dr. N.K. Savant was evaluated by scientists working on rice crop at Panvel, Karjat and Palghar during the *kharif* seasons of 1997 and 1998 (Table 17) (Savant, 2002; Anonymous, 1998).

The following four Practices of Agrotechnology were evaluated:

Practice 1 a. Seedbeds were mixed with black-gray ash obtained by burning of rice hulls @ 1 kg m^{-2}.

Practice 2. Limited green manuring of *gliricidia* leaves was used (~ 3 t ha^{-1}).

Practice 3. Controlled transplanting of rice seedlings (25 hills m^{-2}) was done.

Practice 4. Deep placement of Urea-DAP briquettes @ one briquette of 2.7g for every 4 rice hills was followed.

Highlights :

- Saving of 25 percent seedlings as compared to the traditional method of transplanting.
- Saving of 40 percent N and P and 100 percent saving of K compared to the recommended fertilizers.
- Increased productivity of rice from 11 to 22 percent.
- Economically SIRA generates higher profit than the traditional method.

Table 17. Evaluation of SIRA on the University research farms during *kharif* 1997 and 1998

Sr. No.	Treatment	No. of hills m^{-2}	Grain yield (q ha^{-1}) Two-year average		
			Panvel (Panvel-1) (1997)	Karjat (Sahyadri) (1997)	Palghar (1998)
1	Traditional method	33	35.2	48.9	15
2	20 x 15cm transplanting + 100% recommended fertilizers *	33	43.5 (2214)	62.1	35.9
3	20 x 15cm transplanting + 100% recommended fertilizers	25	42.6 (793)	60.5	27.0
4	Controlled transplanting + Use of Urea-DAP briquettes (Practices 3 and 4)	25	47.9 (3227)	61.5	30.8
5	Rice husk ash + Controlled transplanting + Use of Urea-DAP briquettes (Practices 1a, 3 and 4)	25	47.6 (3270)	71.7	**
6	Rice husk ash + Limited green manuring + Controlled transplanting + Use of Urea-DAP briquettes (Practices 1a, 2, 3 and 4)	25	48.2 (3544)	75.1	42.9
		C.D. (0.05)	4.10	3.02	8.87

* Recommended fertilizer dose : 100 kg +50 kg P_2O_5+ 50 kg K_2O ha^{-1}

** No treatment was not used.

Figures. in bracket indicate net profit in Rs. ha^{-1s}.

Recommendation :

Use of rice hull ash @ 1.0 kg m^{-2} for raising seedlings, application of ~ 3 t ha^{-1} green manuring through *gliricidia* and deep placement (7 to 10 cm) of Urea-DAP (*Vikram*-2) briquettes @ one for four hills of modified (controlled) transplanting method are recommended for higher rice production, increasing fertilizer use efficiency, saving of inputs, sustaining soil productivity and conservation of existing forest.

8.9.1.1 Agriculture Research Station (ARS) Phondaghat, Dist. Sindhudurg

The SIRA has been evaluated on the research farm of ARS, Phondaghat during *kharif* seasons of 1999 to 2002 by using hybrid rice, Sahyadri for its yield and economical benefits. Data collected by using following practices are summarised in Table 18 (Savant, 2002; Anonymous, 2002).

Table 18. Evaluation of SIRA with Karjat hybrid rice-Sahyadri at ARS, Phondaghat, Dist. Sindhudurg (1999 to 2001)

Sr. No.	Treatment	Three year average yield (q ha^{-1})		Cost of cultivation (Rs. ha^{-1})	Monetory returns (Rs. ha^{-1})	Loss(-) or profit (+) (Rs. ha^{-1})	Benefit / cost ratio
		Grain	Straw				
1	15 x 20 cm transplanting (33 hills m^{-2}) + no fertilizer use	24.32	25.00	26819	16592	-10227	0.62
2	15 x 20 cm transplanting (33 hills m^{-2}) + 100 kg N + 50 kg P_2O_5 ha^{-1}	49.13	50.00	29125	33478	+ 4353	1.15
3	Controlled transplanting (25 hills m^{-2} + use of Urea-DAP briquettes (practices 3 and 4)	50.37	51.00	28349	34302	+ 5953	1.21
4	Rice hull ash + controlled transplanting + Urea-DAP briquettes use (Practices 1a, 3 and 4)	52.66	53.00	28636	35836	+7200	1.25
5	Rice hull ash + green manuring (10 t ha^{-1}) + controlled transplanting + Urea-DAP briquettes use (Practices 1a, 2, 3 and 4)	61.28	62.00	29636	41560	+11924	1.40

Practice 1 a. Rice hull ash @ 1 kg m^{-2} was used in seedbed.

Practice 3. *Controlled transplanting* @ 25 hills m^{-2}

Practice 4. Deep placement of Urea-DAP briquettes at 7 to 10 cm soil depth (@ one briquette per 4 square hills).

Note : In treatment 5 (Table 13) *gliricidia* green manuring was used @ 10 t ha^{-1}. It was about three times more than that of Practice 2. This quantity is not affordable for small farmers. It may result in increasing emission of methane gas due to its anaerobic decomposition in soil thus leading to environmental (air) pollution. Therefore, rice yield of treatment 5 was not considered.

Highlights :

- Practices 1a, 3 and 4 saved 25 percent seed and 40 percent fertilizer and increased rice yields.
- Rice farming using SIRA was economical with a profit of Rs. 7200/- ha^{-1}.

8.9.1.2 Regional Agriculture Research Station (RARS), Karjat

Four large scale field trials as detailed below were carried out during *kharif* 2001 on hybrid rice Sahyadri and Jaya variety of rice to demonstrate effectiveness of SIRA on yield of rice at RARS Karjat (Table 19) (Savant, 2002; Anonymous, 2001 k).

i. Sahyadri hybrid rice : Transplanting of rice seedlings at 15 x 20 cm (33 hills m^{-2}) spacing + 50 kg N + 50 kg P_2O_5 + 50 K_2O ha^{-1} applied at transplanting and remaining 50 kg N ha^{-1} one month after transplanting.

ii. Sahyadri hybrid Rice + SIRA

iii. Jaya rice variety and transplanting and fertilizer application as per treatment 1

iv . Jaya rice variety + SIRA

The 'SIRA' in the above experiment includes following four practices.

Practice 1. Application of rice hull ash (@ 1 kg m^{-2}) in seedbed.

Practice 2. *Gliricidia* green manuring (@ 30 q ha^{-1}) before puddling.

Practice 3. Controlled transplanting of rice seedlings @ 25 hills m^{-2}.

Practice 4. Placement of Urea-DAP briquettes (one briquette per four hill square) at 7 to 10 cm soil depth immediately after transplanting.

Highlights :

- There was saving of 25 percent seeds in both the varieties.
- The SIRA produced better crop growth, no. of productive tillers per hill and no. of grains in each panicle than those by the traditional method.
- There was an overall saving of 40 percent N and P and 100 percent K fertilizers with additional yield of 6.7 qha^{-1} of Sahyadri and 3.1 qha^{-1} of Jaya variety.
- Rice growers visiting these demonstration plots were greatly impressed by SIRA and expressed their desire to follow it on their farms.

Table 19. Demonstration trials of SIRA at Regional Agril. Research Station farm Karjat, *kharif* 2001

Rice cultivar	Plot no.	Productive tillers $hill^{-1}$	Height of the plant (cm)	No. of grains $panicle^{-1}$	Yield (q ha^{-1})
Sahyadri	1	12	120	213	69.5
	2	18	133	232	76.2
Jaya	3	9	116	169	57.2
	4	12	123	234	60.3

8.9.1.3 Khar Land Research Station, Panvel

The SIRA has been demonstrated on khar land during *kharif* 1997 to 1999 on Panvel-1 rice variety. Following four treatments were used (Table 20) (Savant, 2002; Anonymous, 1999).

1. Rahu sowing (seeds soaked in water for 24 hrs., then kept in a bag for sprouting, and sprouted seeds are broadcasted

on field) about 50 seeds m^{2} and 150 kg N + 50 kg P_2O_5 + 50 kg K_2O ha^{-1} recommended dose of fertilizers.

2. Traditional row transplanting at 20 x 15 cm spacing, (33 hills m^{-2}) + 100 kg N + 50 kg P_2O_5 + 50 kg K_2O ha^{-1}(recommended fertilizer use)
3. Controlled transplanting (25 hills m^{-2}) + 100 kg N + 50 kg P_2O_5 + 50 kg K_2O ha^{-1} (recommended fertilizer dose)
3. Controlled transplanting (25 hills m^{-2}) + Use of Urea-DAP briquettes (58 kg N ha^{-1} + 29 kg P_2O_5 ha^{-1}) deep placement of one briquette weighing 2.7 g per four hills + Limited *gliricidia* green manuring @ 30 q ha^{-1} (Practices 2,3 and 4).

Table 20. Demonstration trial of SIRA at Khar Land Research Station, Panvel, 1997-1999

Sr. No.	Method of rice cultivation	Seedlings or hills m^{-2}	Rice yield (q ha^{-1})		
			1997	1998	1999
1	Rahu sowing + recommended fertilizer	50	56(a)	39(a)	48(a)
2	Traditional transplanting (20 x 15cm) + recommended fertilizer	33	58(a)	34(a)	46(a)
3	Controlled transplanting + recommended fertilizer	25	54(a)	41(b)	48(a)
4	SIRA, Practices 2, 3 and 4	25	61(a)	46(b)	54(b)

Note : Every year, yields followed by the same letter are considered statistically on par with each other.

Highlights :

During 1997, all the treatments were statistically on par with each other, while SIRA (Practices 2, 3 and 4) during 1998 and 1999 were superior to rest of the treatments. It was observed that SIRA required less seeds and fertilizers and recorded 12 to 35 percent increase in the grain yield over the other treatments (Yadav *et al.*, 2002).

8.9.2 Mahatma Phule Krishi Vidyapeeth, Rahuri, Dist. Ahmednagar

i. Agricultural Research Station (ARS), Igatpuri, Dist. Nasik

In order to assess the potential of SIRA in comparison to traditional rice cultivation, an experiment was conducted at ARS, Igatpuri by adopting Practices 3 and 4 only (Tables 21 and 22) (Savant, 2002; Anonymous, 2001 l) .

Table 21. Evaluation of performance of controlled transplanting (Practice 3) and Urea-DAP briquettes (Practice 4) on rice

Method of cultivation	Productive tillers m^{-2} (Nos.)	Weight of panicle (g)	Grain filling (%)	Yield (q ha^{-1})	Yield increase (%)
1. Traditional method of transplanting + Recommended fertilizers i.e. 100 kg N + 50 kg P_2O_5 + 50 kg K_2O ha^{-1}	172	2.18	78	31.6	-
2. Controlled transplanting (Practice 3) + 56 kg N + 29 kg P_2O_5 ha^{-1} using Urea-DAP briquettes (Practice 4)	268	2.28	83	43.1	36.4

Highlights :

- By adopting Practices 3 and 4, there was 40 percent saving in the use of fertilizers than the traditional method.
- Effective use of briquettes resulted in increase in productive tillers m^{-2}, weight of the panicles and percentage of grain filling.
- Controlled transplanting and use of Urea-DAP briquettes recorded increased yield of rice by 36.4 percent.

Table 22. Benefit : cost ratio due to use of SIRA Practice 3 and 4

Method of cultivation	Yield (q ha^{-1})	Cost of cultivation (Rs. ha^{-1})	Income (Rs. ha^{-1})	Net profit (Rs. ha^{-1})	Benefit / cost ratio
1. Traditional method of transplanting + Recommended fertilizers i.e. 100 kg N + 50 kg P_2O_5 + 50 kg K_2O ha $^{-1}$	31.6	12975	18960	5985	1.46
2. Controlled transplanting (Practice 3) + 56 kg N + 29 kg P_2O_5 ha $^{-1}$ using Urea-DAP briquettes (Practice 4)	43.1	12350	25860	13510	2.09

- Effective use of briquettes by deep placement after transplanting, saved the fertilizers to the extent of Rs. 625/- ha^{-1} than the traditional method.
- The increase in the yield resulted in the increase of net profit, and the benefit : cost ratio.
- Rice scientists have concluded that farmers should adopt SIRA for increasing rice production (Savant, 2005).

Agricultural Research Station, Igatpuri (Nasik District) conducted rice trials on farmers' fields in Nasik district during the *kharif* 2000 and 2001 (Anonymous, 2001 m).

The following three methods of rice cultivation have been used as shown in Table 23.

Treatment 1 : Local method of cultivation having uncontrolled and irregular transplanting along with traditional method of broadcasting of fertilizers.

Treatment 2 : Recommended method i.e. transplanting of seedling at 15 x 20 cm, (33 hills m^{-2}) + application at 100 kg N + 50 kg P_2O_5 and 50 kg K_2O ha^{-1}. Full dose of P and K was applied before transplanting and 100 kg N ha^{-1} applied in 3 split doses; 40 kg N at transplanting, 40 kg at one month after transplanting and 20 kg N at flowering stage.

Treatment 3 : In SIRA, Practices 3 and 4 were only used; controlled transplanting (25 hills m^{-1}) and application of 58 kg N + 29 kg P_2O_5 ha^{-1} through briquettes immediately after transplanting; one briquette of 2.7 g each in 4 hills placed at 7 to 10 cm deep in soil.

Highlights :

- Adoption of Practices 3 and 4 alone resulted in 37 percent (in the year 2000) and 78 percent (in 2001) increase in grain yield and 40 percent (in 2000) and 70 percent (in 2001) straw yield compared to local method.
- Compared with recommended method of rice cultivation, SIRA produced 6 percent (in2000) and 29 percent (in 2001) higher grain yield and 5 percent (in 2000) and 25 percent (in 2001) straw yields.
- In short, adoption of SIRA (Practices 3 and 4) not only saved on seeds and fertilizers but also increased grain and straw yields.
- During the *kharif* 2000 season, 15 trials and in *kharif* 2001 season, 11 trials were conducted on the farmer's fields in Nasik district (Table 23) (Savant, 2002 ; Anonymous, 2001 m) on transplanted rice. Rice grain and straw yields increased in controlled transplanting with use of Urea-DAP briquettes as compared to farmers' method and recommended method of transplanting and fertilizer level application method (Savant, 2002 ; Anonymous, 2001 m).

Table 23. Rice production in local, recommended and SIRA (Practices 3 and 4 only) on farmers' fields, Nasik District, *kharif* 2000 and 2001

Tahsil	No. of farm trials	Rice cultivation method					
		Local method		Recommended method of transplanting & fertilizer use		Controlled transplanting & use of Urea-DAP briquettes	
		Grain ($q\ ha^{-1}$)	Straw ($q\ ha^{-1}$)	Grain ($q\ ha^{-1}$)	Straw ($q\ ha^{-1}$)	Grain ($q\ ha^{-1}$)	Straw ($q\ ha^{-1}$)
***Kharif* 2000**							
Igatpuri	3	28.2	31.2	32.2	38.7	36.2	43.5
Rajur	3	26.8	32.1	31.9	38.2	33.3	39.9
Peth	3	25.7	30.6	34.9	41.9	36.3	43.6
Trimbakeshwar	3	25.2	28.7	34.8	41.6	36.8	42.9
Dindori	3	24.4	29.3	34.7	41.6	36.2	43.4
Average		26.1	30.4	33.7	40.4	35.7	42.7
***Kharif* 2001**							
Igatpuri	2	26.6	31.9	32.4	38.9	35.4	42.5
Peth	3	26.8	32.1	37.9	45.5	44.9	53.7
Trimbakeshwar	3	27.1	32.5	40.3	48.7	47.5	57.2
Dindori	3	26.8	32.1	39.8	47.7	61.7	74.0
Average		26.9	38.1	38.1	45.8	48.2	57.1

ii. Agricultural Extension Wing, College of Agriculture, Pune, *kharif* 2001

During 2001 *kharif* season, 66 students undergoing Rural Agricultural Works Experience Programme (RAWE) from College of Agriculture, Pune have carried out 28 trials of SIRA on farmers' fields (15 using Practices 2, 3 and 4 and 13 using Practices 3 and 4) in nine villages of Bhor Tahsil (Natambi, Karanje, Ambawade, Nerle, Khanapur, Sangavi, etc.) of Pune district. In all the trials of SIRA, an average increase in grain yield of 11.5 q ha^{-1} (31%) over the traditional method was noticed (Savant, 2002 ; Anonymous, 2001 n).

Chapter 9

Integrated use of Urea-DAP Briquettes for Upland (drilled) Rice

---

9.1 Integrated use of Urea-DAP briquettes for upland (drilled) rice

Practice 1. Improved / hybrid drilled or dibbled paddy seeds in paired-rows

- **Variety** : Use recommended variety of drilled paddy for western Maharashtra and Marathwada. For different parts of India, use recommended rice varieties of the region.
- **Seeds** : Use seeds 10-20 percent less than the recommendation.
- Sow the treated seeds in paired-rows after land preparation and before the rains or after the rains at soil field capacity conditions. Considering the soil slope / relief, seeds are sown preferably in south-north direction in paired-row.
    (a) The guide should first prepare paired-rows at 15 cm (6") distance and in every paired-row keep 30 cm (12") walking lane (Fig. 40) (Savant, 2007).
    (b) Small farmers should perform sowing of rice in paired-rows by dibbling.
    (c) Seeds are sown by newly developed paired-row iron seed drill or usual wooden seed drill (15 cm distance in every two lines) (6") using 30 cm (12") distance in every paired-row i.e. by improved paired-row seed drill (Photo 17) (Savant, 2007).

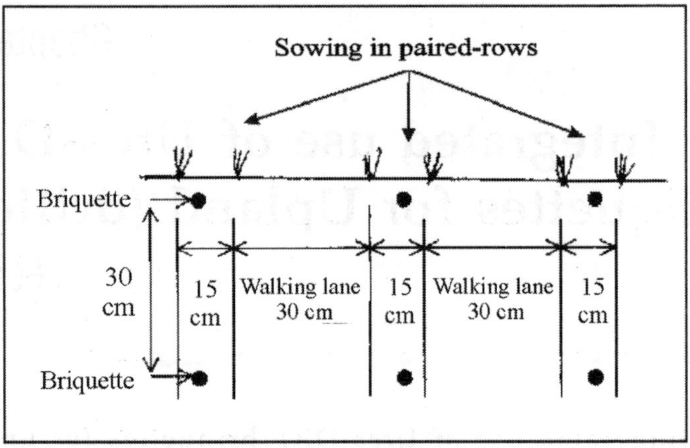

Fig. 40. Sowing/ dibbing of paddy seeds in paired rows

Practice 2. Use of limited and efficient organic manure in paired-rows only

Practice 2a. Before germination of rice seeds

Photo 17. Paired-row seed drill (15 cm spacing with a walking lane of 30 cm)

- Apply a mixture of rice hull black ash (or *Bhatura* 1, which is organic manure available in market) 10% + well decomposed FYM (or compost prepared from rice straw) 90%, (5 to 8 baskets per R i.e. 5-8 t ha^{-1}) in only paired-rows (after sowing of seeds).

NB : In case potash is low in soil as per soil testing, apply 10 to 20 kg K_2O ha^{-1} through muriate of potash (KCl) in paired-rows by mixing with *Bhatura*-1 or FYM.

Practice 2 b. After germination of paddy seeds

- Spread *gliricidia* green leaves (approx. 30 kg per R i.e. about 3 tha^{-1}) in paired-rows only.

Benefits : This will help in saving of water, increasing organic matter and nutrients in soil and will help in minimizing the weed problem.

Practice 3. Efficient use of Urea-DAP briquettes after germination of paddy seeds (after 10 days) in paired-rows

Apply one Urea-DAP briquette by deep-placing (4-7 cm) in soil at a distance of 30 cm by hand. After placement of briquettes in soil if hole remains open, close it by hand. If necessary irrigate once. The crop should receive 68 kg N and 34 kg P_2O_5 ha^{-1}. Thus, there will be a saving of 25 percent of chemical fertilizers as compared to the recommended fertilizer dose.

Practice 4. Water management

- Considering the proportion of rains in *kharif* season, apply 3 to 4 irrigations of 50-100 mm depth particularly at placement of briquettes in soil, at tillering, flowering and grain filling stages of rice.

NB : In case there is pest / disease problem on rice, follow integrated pest / disease management.

9.2 Results of field demonstration trials on drilled rice

Farmers like Shri. Patil (Tah. Khadarwadi, Dist. Belgaon) in *kharif* season of 2001 and Shri. Dhanavade (Mudshingi, Tah. Karveer, Dist. Kolhapur) in *kharif* season of 2002, used Integrated (*Charsutri*) agrotechnology on their fields and obtained increased drilled paddy yields.

- Shri. Sanjay Tiravade and Shri. Shivaji Chaugule, farmers from Radhanagari (Dist. Kolhapur) used this agrotechnology in *kharif* season of 2003 on their fields for the first time. They obtained increase in yields of drilled paddy. Shri. Ganesh Ghorpade, Agricultural Development Officer and Shri. Magdum, Project Officer from Z.P. Kolhapur expressed satisfaction after visiting these trials (Anonymous, 2003).

- Zilla Parishad, Kolhapur conducted field demonstration trials with the help of farmers in Bawda, Chandgad, Panhala, Karveer, Gadhinglaj and Bhudargarh tahsils on Integrated (*Charsutri)* drilled paddy farming agrotechnology (Photo 18). Twenty trials were conducted using Jaya, Sahyadri, Ratnagiri 1, Radhanagari 185-2, Sonam, etc. rice varieties (Anonymous, 2003). For comparison, Farmers' method, Recommended method and Integrated (*Charsutri*) agrotechnology method were used. In Integrated (*Charsutri*) agrotechnology, the requirement of seeds and fertilizers were less, and drilled paddy yields increased by 27 percent over Farmers' method and by 11 percent over Recommended method (Photo 18) (Anonymous, 2003).

Photo 18. Dr. Savant and Shri. Magdum, Agriculture Development Officer, Z.P. Kolhapur observing *Charsutri* drilled paddy trial on farmers' field (Shri. D.S. Chaugule, Panhala, Dist. Kolhapur)

Farmers' response

Good quality of Urea-DAP briquettes should be made available to the farmers in time. Similarly Department of Agricultural Extension of the Agricultural Universities should organise farmers' rallies and provide training to the farmers regarding the use of briquettes.

Chapter 10

Integrated use of Urea-DAP Briquettes for Wheat

Practice 1. Paired-row sowing or dibbling (hand placement) of improved / hybrid wheat cultivars after seed treatment

Use improved wooden or iron four coulter seed drill with 15 cm (6") distance in every two coulters and in every two paired-rows a walking lane of 30 cm (12") distance. Considering the soil slope the seeds are sown in south-north direction. (Fig. 41; Photo 19) (Savant, 2007).

Seed rate : 800 to 900 g seeds (with seed treatment) per R i.e. 80 to 90 kg ha^{-1}.

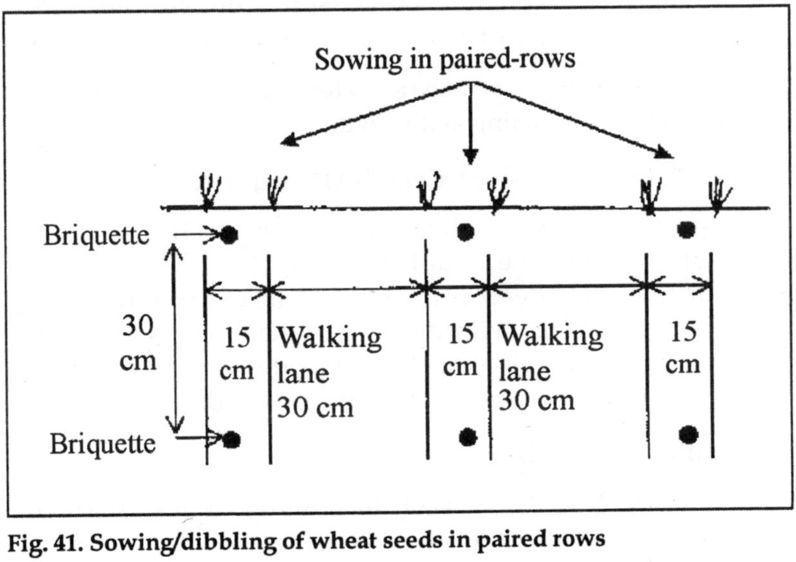

Fig. 41. Sowing/dibbling of wheat seeds in paired rows

Photo 19. A farmer Shri. Punjaram Jadhav of Bilda village, Pah. Phulambri, Dist. Aurangabad observing his field with wheat crop grown in paired-rows

Practice 2. Limited and efficient use of organic manures

- After sowing apply minimum 4-5 baskets of compost per R (0.01 ha) i.e. minimum 4-5 t ha^{-1} of compost prepared from wheat straw or well decomposed FYM / organic manure only in paired-rows by mixing with soil.
- Similarly after sowing of wheat (after wheat germination) use agroforestry approach i.e. spreading *gliricidia* green leaves (approx. 3 t ha^{-1}) in paired-rows.

NB : If necessary apply potash (K_2O) fertilizer as per soil analysis in paired-rows by mixing with organic manure.

Practice 3. Efficient use of Urea-DAP briquettes

After germination of wheat seeds (about 10 days after germination) apply one Urea-DAP briquette of 2.7 g weight, 5 to 7 cm (2 to 3″) deep into the soil in every paired-row approximately 30 cm (12″) distance by hand or by applicator (Vijay 2). After deep placement of briquettes if the holes remain open at insertion sites, close the same by hand.

Application of fertilizers :

Apply 68 kg N + 34 kg P_2O_5 ha^{-1} which is less than the recommended fertilizer level.

Table 24. Demonstration trials on integrated (*Charsutri*) wheat, on farmers' fields, *rabi* 2002-03 to 2006-07

Institution/ Farmer	*Rabi* season	No. of trials	Cultivar	Place	Average production (q ha^{-1})		Increase in production %	Farmers Opinion
					Farmers' method	*Charsutri* method		
Shri. Nanasaheb Deshmukh	2002-03	1	HD 2189	Aurangabad	*	*	*	Production increased due to use of briquette
Shri. Jedhe, Shri Bhumkar	2003-04	5	HD 2189	Pune, Aurangabad	16.2 (24)	24.0 (35.6)	48	
Maharashtra state Agriculture Depart., Pune	2004-05	17	HD 2189	Tah. Phulambri, Dist. Aurangabad	14.5	20.8	43	At some places Vijay-2 used
Agriculture School, Pune & Janaseva Foundation, Pune	2005-06	13	HD 2189, Trimbak	Tah. Bhor, Dist. Pune	30.1 (50.1)	37.5 (61.5)	24	-
Agriculture School, Pune	2006-07	4	Trimbak	Bhor, Dist. Pune	31.4	36.3	16	-
Janaseva Foundation, Pune	2006-07	22	Lokwan	Pampasarovar Dang (Gujrat state)	17.2	28.8	67	-
	2006-07	8	Junagad, HD 2189	Surat	11.5	17.2	49	-
Casp Plan, Pune	2006-07	1	Trimbak	Tah.Maval, Dist. Pune	-	35.0	Appox.30	

Krishi Vigyan Kendra, Baramati	2006-07	3	Trimbak	Tah. Baramati, Dist. Pune	30.8	39.7	29	
Dept. of Agriculture, Maharashtra, Pune Division	2007-08	34	HD 2189, NIAW-34	Kopargaon, Sakur, Akole, Dist. Ahmednagar	38.0	47.7	25	
Shri, Ingavale, Agriculture College, Pune	2007-08	5	HD 2189, Trimbak	Nasrapur, Dist. Pune	38.0	44.4	15	
Janseva Foundation, Pune	2007-08	42	Trimbak	Dang Dist., Gujarat	35.0	53.0	51	

Figures in bracket indicate straw yields

Practice 4. Water management (as per the need)

Apply at least 4 irrigations of 100 mm depth with proper layouts as per the requirement (after every three weeks).

NB : In case there is incidence of pests and diseases, follow integrated pest / disease management.

10.1 Results of demonstration trials on wheat conducted on the farmers' fields

Janaseva Foundation, Pune conducted 22 field demonstration trials on integrated (*Charsutri*) wheat at Pampasarovar in Dang district and in Songadh, Vyara and Uchhal Tahsils of Surat district in Gujarat State. Every trained farmer conducted Integrated (*Charsutri*) method and Farmers' method of wheat cultivation each on 5 R (0.05ha) land. Wheat was sown in south-north direction in paired-rows. Other practices of wheat cultivation were the same in both the methods (Table 24) (Anonymous, 2005a ; 2005 b; Anonymous, 2006 a; Anonymous, 2007 a ; 2007 b ; 2007 c; 2007 d; Savant, 2007).

At Pampasarovar, the average wheat yields were 17.2 qha^{-1} in Farmers' method and 28.8 qha^{-1} in Integrated (*Charsutri)* method (67 percent more yield) by the tribal farmers. In Surat district, yield data were obtained from 8 farmers. In Farmers' method, the average wheat yield was 11.5 qha^{-1}, while in Integrated (*Charsutri)* method it was 17.2 qha^{-1} (49 percent more). Thus, Integrated (*Charsutri*) method of wheat cultivation was found to be profitable.

The CASP-Plan, Kanhephata, Maval tahsil in Pune district and Krishi Vigyan Kendra, Baramati in Pune district conducted Integrated (*Charsutri*) wheat cultivation trials on farmers' fields. In these trials, the farmers obtained increase in wheat yields by 35% and 39%, respectively. Both the organisations informed that Integrated (*Charsutri)* method of wheat cultivation was very simple, less expensive and resulting in more crop yields to the small farmers.

Farmers' response

Integrated (*Charsutri*) wheat cultivation has been found profitable. However, training is required regarding Integrated (*Charsutri*) method for the farmers.

- ***Rabi* season 2003-04** : In all 5 trials (four in Pune and one in Aurangabad districts) were conducted during the *rabi* 2003-04 on wheat. Crop growth was improved. The farmers used Integrated (*Charsutri*) agrotechnology in which 35 percent less fertilizers were applied. The farmers obtained on an average 48 percent more wheat yields and about Rs.6000/- ha^{-1} profit.
- ***Rabi* season 2004-05** : In all 17 trials were conducted on farmers' fields in Bilda village of Phulambri tahsil of Aurangabad district during the *rabi* 2004-05. In Integrated (*Charsutri*) method the average wheat yields obtained was 20.8 qha^{-1} while in usual Farmers' method it was 14.5 qha^{-1} wheat yield.
- ***Rabi* season 2005-06** : In Integrated (*Charsutri*) method of wheat cultivation sowing was done in south-north direction so as to get enough sunlight and air. Due to deep placement of Urea-DAP briquettes, wheat crop growth was good and the crop was green in colour up to harvesting. The number of tillers and panicles were 33 percent more in Integrated (*Charsutri*) method than the usual Farmers' method. In some field trials, deep placement of briquettes were delayed by about 30 days, which resulted in lowering of wheat yields to a small extent. It was observed that Integrated (*Charsutri*) method recorded increase in yields on the fields of small farmers when compared with usual Farmers' method. One farmer used gobar gas slurry which increased Integrated (*Charsutri*) wheat yields by 37 percent. In Integrated (*Charsutri*) method, wheat yields were 37.5 qha^{-1} grain and 61.5 q ha^{-1} straw. The Farmers' method recorded 30.1 q ha^{-1} grain and 50.1 qha^{-1} straw yields of wheat. In Integrated (*Charsutri*) method, wheat yields increased on an average by 24 percent in grain and 22 percent in straw (Photo 20).

Photo 20. Integrated (*Charsutri*) wheat trial on the filed of Mrs. Dwarkabai Salavi, Tah. Bhor, Dist. Pune, *rabi* 2005-06

- ***Rabi* season 2006-07** : Six farmers in Bhor Tahsil (Dist. Pune) used Integrated (*Charsutri)* method for wheat cultivation each on 2 ha field. For comparison wheat crop was also grown with Farmers' method. Sowing was done in south-north direction. Integrated (*Charsutri*) method resulted in increasing wheat yields up to 30 percent. In two trials, wheat yields obtained were less due to incidence of disease. Since yields obtained were higher for the last two years, the farmers decided to go in for Integrated (*Charsutri)* method of wheat cultivation (Photo 21).

Photo 21. Farmers from Bhor (Dist. Pune) used iron paired-row seed drill (15 cm distance) and successfully sown *Charsutri* wheat, *rabi* 2006-07

- ***Rabi* season of 2007- 08 :**

Agriculture Department of Maharashtra (Pune Region) organised 325 wheat demonstration trials (highest number in the State) on the farmers' fields covering 45 villages in Malshiras tahsil, dist. Solapur. Shri. Abasaheb Dhapte, Tahsil Agriculture Officer and his staff first organised training programmes for the farmers on Integrated (*Charsutri)* wheat agrotechnology in 20 villages. They provided Urea-DAP briquettes to the farmers @ Rs. 10.30 kg^{-1} although the official rate was Rs. 12.75 kg^{-1} and fabricated 15 improved seed drills at local level for the use of farmers. Shri. Dhairyasheel Mohite Patil, Chairman of Panchayat Samitee, provided good quality of wheat seeds. Farmers conducted wheat trials as per Integrated (*Charsutri)* method. Due to use of Urea-DAP briquettes, there was saving of expenditure on fertilizers by about Rs. 1300/- ha^{-1}. Due to paired-row sowing of wheat with improved seed drill in south-north direction, there was enough sunlight for the crop to carry out photosynthesis activity and less incidence of disease. The crop growth was excellent with more number of tillers. The grain formation was good with more number of grains per plant. It is expected that the yields will be increased by about 20 percent due to Integrated (*Charsutri)* method as compared to usual Farmers' method with use of less fertilizers (Photos 22, 23 and 24) (Agro One, 2008 b).

Photo 22. Paired-row sowing of *Charsutri* wheat at Malshiras, dist. Solapur, *rabi* 2007-08

Photo 23. Deep placement of Urea-briquettes by women workers of *Charsutri* wheat, at Malshiras, Dist. Solapur, *rabi* 2007-08

Photo 24. Sowing in progress with locally manufactured paired-row seed drill

Shri Balasaheb Meher, Narayangaon Mandal Agriculture Officer, and his staff organised Integrated (*Charsutri)* wheat demonstration trials on farmers' fields in 20 villages of Junnar tahsil in Nasik district. Out of these, a model demonstration trial was conducted at Warulwadi village on the field of Shri. Shivdatta Sante. All the trials were conducted under Cereal Development Programme, National Food Security Programme and Water Sources Improvement Project. They organised farmers' rallies, group discussions, visit to the fields to spread awareness among the farmers about the Integrated (*Charsutri)* wheat agrotechnology. They also showed CD about Integrated (*Charsutri*) method of wheat cultivation to the farmers. All the farmers followed Integrated (*Charsutri)* method of wheat cultivation. For a model trial on the field of Shri. Sante a seed drill was fabricated to make markings, sowing and covering with soil operations simultaneously. An improved variety of wheat (cv. Trimbak) was used by the farmers. Due to use of Urea-DAP briquettes, there was saving on fertilizer by about 20 to 30 percent. For Integrated (*Charsutri)* method the total expenditure was Rs. 12500/- ha^{-1} including land preparation, seeds, seed treatment, sowing, weeding, Urea-DAP briquettes, insecticide, harvesting and threshing. The expenditure on Farmers' method was about Rs. 20000/-ha^{-1}. Integrated (*Charsutri)* method recorded about 20 percent increase in wheat yields. A model trial recorded 55 qha^{-1} wheat yield (about 40 percent increase) (Photos 25 and 26) (Agro One, 2008 c).

Photo 25. Shri. Devdatta Sante, a farmer with his excellent *Charsutri* wheat trial at Waruladi, Dist. Pune, *rabi* 2007-08

Photo 26. Shri. K.V. Deshmukh, Joint Director of Agriculture discussing with the farmers regarding paired-rows seed drill for *Charsutri* wheat sowing at Warulwadi, Dist. Pune, *rabi* 2007-08

Shri. K.V. Deshmukh, Joint Director of Agriculture, Pune region organized 34 demonstration trials on the farmers' fields on *Charsutri* (Integrated organic-inorganic farming) wheat during *rabi* 2007-08 at Kopargaon, Sakur and Akole in Ahmednagar district. For each trial, there were two plots viz., *Charsutri* method and Farmers' method of wheat cultivation for comparison. In *Charsutri* method (see Chapter 8) farmers observed less weed problem with a saving of expenditure on fertilizers by about 40% as compared to the Farmers' method. *Charsutri* method recorded on an average 47.7 qha^{-1} of wheat yield as compared to 38.0 qha^{-1} in Farmers' method. Thus, *Charsutri* method resulted in 25% increase in wheat yields with 40% saving of expenditure on fertilizer use (Agro One, 2008 d).

***Rabi* 2008-09 :**

In Ahmednagar district, Shri. Vikas Patil, Superintending Agriculture Officer, organized 54 demonstration trials on wheat on farmers' fields in Sangamner, Kopargaon, Rahata and Akole tahsils during the *rabi* 2008-09. These trials were conducted to compare the Farmers' method (traditional method) and *Charsutri* i.e. Savant's Integrated Wheat Agrotechnology (SIWA) method of wheat cultivation using NIAW 301, NIAW 34, HD 2189, 2496 and Lok One wheat cultivars (Anonymous, 2009 c).

The results showed that majority of the farmers (35 out of 54 farmers) were in the lower productivity class of 11 - 20.9 qha^{-1} with a mean yield of 15.5 qha^{-1}. These farmers were benefited more increasing their mean yield to 19.4 qha^{-1} (25% increase). (Table 25). The over all increase in yields of all the farmers were by about 20%. This increase in yield of wheat was attributed to increase in number of ears per plant, ear length and 1000 grain weight (Anonymous, 2009 c). The calculated 't' value was highly significant indicating that SIWA method was significantly superior to the Farmers' method in increasing the wheat yields (Table 25).

Table 25. Wheat yields on farmers' fields in Sangamner, Kopargaon, Rahata and Akole tahsils of Ahmednagar district (Maharashtra), *rabi* 2008- 09

Productivity class (qha^{-1})	No. of farmers	Mean wheat yield (qha^{-1})			
		Farmers' method	SIRA	% increase	' t 'cal.
11- 20.9	35	15.5 (2.43)	19.4 (3.62)	25.2	12.97**
21 - 30.9	12	24.8 (2.15)	28.7 (2.73)	15.7	
31 - 30.9	7	35.8 (2.19)	40.4 (3.99)	12.8	
Total	**54**	**20.2 (7.55)**	**24.2 (8.06)**	**19.8**	

Figures in bracket indicate standard deviation.
Wheat cultivars used : NIAW 301, NIAW 34, HD 2189, 2496, Lok One.

Chapter **11**

Integrated use of Urea-DAP Briquettes for Rabi Sorghum (Jowar)

Practice 1. Sowing / dibbling of improved / hybrid sorghum seeds in paired-rows

Sorghum seeds used for sowing should be seed-treated. The seed rate is 90 to 100 g per R i.e. 9 to 10 kg ha^{-1} Sowing is done by paired-row seed drill with 15 cm distance between rows or by dibbling and a walking lane of 30 cm (Fig. 42) (Savant, 2007).

Practice 2. Limited use of organic manure

- Apply compost prepared from sorghum crop residue or FYM or organic manure (minimum 5 to 8 baskets per R (0.01 ha) only in paired-rows by mixing with soil. Considering the soil slope, sowing is done preferably in south-north direction. Use improved paired-row seed drill for sowing. After 10-15 days of sowing carry out thinning.

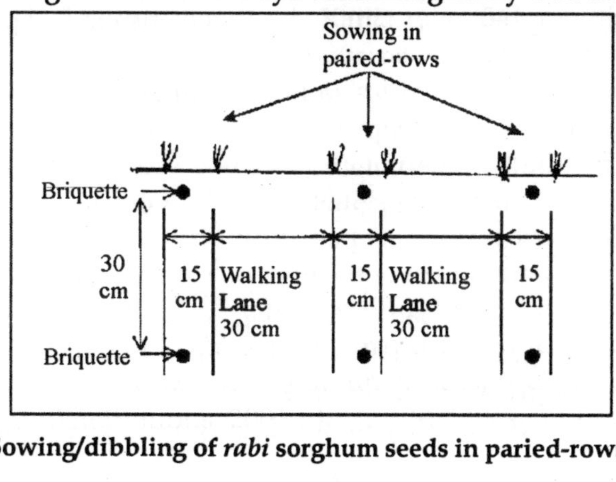

Fig. 42. Sowing/dibbling of *rabi* sorghum seeds in paried-rows

- Using agroforestry approach, spread *gliricidia* green leaves (approx. 30 kg per R i.e. ~ 3 t ha^{-1}) only in paired-rows at thinning.

Practice 3. Efficient use of Urea-DAP briquettes

After seed germination, one Urea-DAP briquette is deep-placed only in paired-rows at 5-7 cm depth in soil at every 30 cm (12") distance. After deep placement of briquettes in soil, if the holes remain open at the sites of placement close the same by hand.

Fertilizer levels per hectare : Apply 68 kg N ha^{-1} + 34 kg P_2O_5 ha^{-1} (30 percent less than the recommended fertilizer dose).

Practice 4. Water management

As per requirement give four irrigations after every three weeks (at deep placement of briquette, tillering, flowering and milk or grain filling stages).

NB : In case there is incidence of pests and diseases in sorghum, use integrated pest-disease management.

11.1 Results of field demonstration trials on *rabi* sorghum (jowar)

- ***Rabi* season 2004-05** : Six trials were conducted on farmers' fields at Mulshi (Dist. Pune) on Integrated (*Charsutri*) *rabi* sorghum cultivation (cv. Maldandi) which resulted in 25 percent higher yields over Farmers' method and 11 percent higher over the general recommended practice (Photos 27 and 28). Similarly one trial was conducted on *rabi* sorghum on farmers' field (Shri. Shendge at Ratranjan, Tah. Barshi, Dist. Solapur) by Dry Farming Research Station, Solapur (Mahatma Phule Krishi Vidyapeeth, Rahuri). The variety used was Phule Yashoda with two irrigations. The farmer obtained 11 percent higher yield in Integrated (*Charsutri*) method over the general recommended method.
- ***Rabi* season 2005-06** : After deep placement of briquettes in soil, sorghum growth was found to be increased. The crop was green until harvest with good grain filling. In Integrated (*Charsutri*) method, sorghum yields of small

farmers were increased (up to 35 percent). In the case of Integrated method (*Charsutri*), sorghum yields were on an average 37.7 qha^{-1} grain and 125 qha^{-1} fodder. While in general recommended practice method average yields were 30.3 qha^{-1} grain and 93.9 qha^{-1} fodder. In Integrated (*Charsutri*) method, sorghum grain and straw yields increased by 33 percent, although less seeds and fertilizers were used (Table 26) (Anonymous, 2006 a ; Anonymous, 2005 c; 2005 d; Anonymous, 2007 e).

Photo 27. Demonstration trial on sorghum, at Mulshi, Dist. Pune, *rabi* 2004-05

Photo 28. Demonstration trials on improved cultivar of sorghum on farmers' field at Mulshi, Dist. Pune, *rabi* 2005-06

Farmers' response :

Use of Urea-DAP briquettes requires more time for deep placement in soil. Reducing this time and labour charges, there is a need for organising farmers rallies and training. The farmers responded that use of Urea-DAP briquettes reduces weed problem.

Table 26. Demonstration trials on integrated (*Charsutri*) sorghum on farmers' fields, *rabi* 2004-05 to 2006-07

Rabi season	Farmer / Institution	Cultivar used / Irrigations given	Average production (q ha^{-1})		Increase in production, %
			Farmers' method	*Charsutri* method	
2004-05	6 farmers, At post Mulshi, Dist. Pune (With the help of Shri. Bhumkar)	Maldandi, 2 to 4 irrigations	22.0 (farmer) 24.7 (recommend ed)	27.3	10 - 24
2004-05	Shri. Shendage Village-Ratanjan, Tah.Barshi (With the help of Agril. Research Centre)	Phule Yashoda, 2 irrigations	29.2 (farmer) 44.2 (recommend ed)	47.0	6 - 60
2005-06	Shri. Maral, Sou. Salvi and other farmers (With the help of Janaseva Foundation, (Selvassa), Pune and Shri.	Maldandi, 2 to 3 irrigations	30.3 (93.9)	37.7 (125)	33
2006-07	Shri. Maral, Shri. Bandal, Sou. Salvi and Shri. Bhorade (With the help of Shri. Ingavale, Shri. Jedhe, Agriculture Extension Wing,	Maldandi, 2 irrigations	19.9 (51.9)	21.8 (54.8)	10

2007-08	Shri. Maral, Shri. Konde, Shri. Bandal, Shri. Salvi (with the help of Shri. Ingavale and Shri. Jedhe, Agricultural Extension Department, College of Agriculture, Pune)	Maldandi, 2 irrigations	14.1 (52)	18.3 (55)	10

Figures in bracket indicate fodder yield.

Chapter **12**

Integrated use of Urea-DAP Briquettes for Pearl Millet (Bajra) (Only for Research)

Practice 1. Sowing/ dibbling of improved or hybrid seeds in paired-rows after seed treatment

Variety : Sraddha, Saburi, etc. improved / hybrid.

Seed rate : 30 to 40 g per R (3 to 4 kg ha^{-1}), while sowing mix seeds with fine sand/well decomposed organic manure in 1:1 proportion.

Seed treatment : Dip the seeds in 20 percent solution of common salt. The seeds at the bottom are dried in shed and then used for sowing.

Sowing : Considering the soil slope, preferably the seed is sown in south-north direction using improved four coulter seed drill. For reducing the cost of cultivation on thinning sowing may be done by hand placement in soil.

Distance : Distance between paired-row is 15 cm with a walking lane of 45 cm (Fig. 43) (Savant, 2007). However, there is a need for carrying out research on walking lane of 30 cm in pearl millet considering the simplicity and convenience of the farmer so that seed drill can be used for sorghum and pearl millet crops.

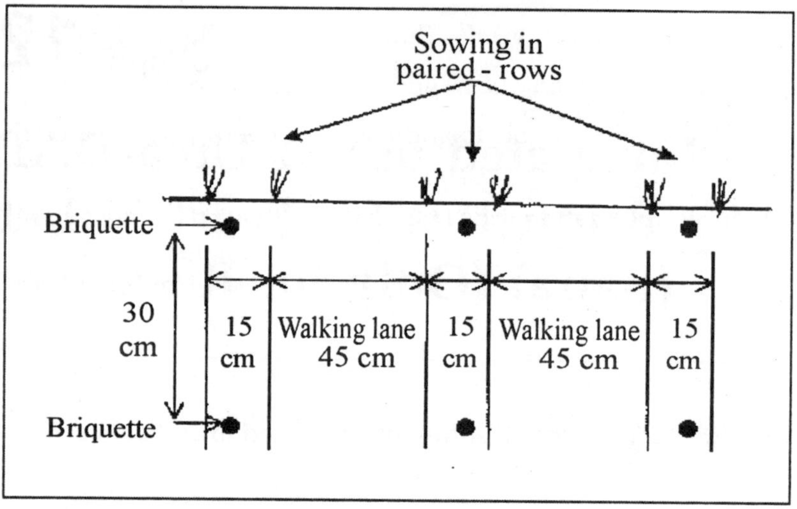

Fig. 43. Sowing/dibbling of pearl millet in paired rows

Practice 2. Limited use of organic manure

- Use well decomposed compost prepared from pearl millet fodder or FYM or any organic manure, minimum 5 to 8 baskets per R (0.01 ha) at sowing only in paired- rows by mixing with soil.
- After seed germination, if possible use agroforestry approach by spreading *gliricidia* green leaves (green manuring) in every paired-row (approximately 30 kg per R i.e. 3 t ha^{-1}).

NB : After soil testing, if necessary apply required quantities of potash fertilizer (muriate of potash) by mixing with organic manure only in paired-rows.

Practice 3. Deep placement of Urea-DAP briquettes

After pearl millet germination (after 10 to 15 days), place one briquette at 5 to 7 cm (2″ to 3″) depth in soil by hand / implement in between paired-rows at 30 cm (12″) distance. If the holes remain open at the sites of placement close the same by hand.

Fertilizer levels : Apply 150 kg briquettes ha^{-1} (52 kg N + 26 kg P_2O_5 ha^{-1}).

Practice 4. Water management

Considering the climate and soil properties only 2-3 irrigations of 100 mm Depth (at sowing / seed placement, 4-5 days after sowing, at briquette placement, tillering, boot and milk or grain filling stages) may be applied.

NB : In case there is incidence of pests/diseases, follow integrated pest-disease management.

12.1 Results of field demonstration trials on pearl millet, *kharif* 2006

Pearl Millet Research Scheme at College of Agriculture, Dhule (Mahatma Phule Krishi Vidyapeeth, Rahuri) conducted integrated (*Charsutri*) pearl millet trial during the *kharif* 2006. They used all the four practices / steps (paired-row sowing, limited use of organic manure in paired-rows, deep placement of Urea-DAP briquettes in paired-rows only at 30 cm distance, water management as required). Only *gliricidia* green leaves were not applied along with organic manure. The variety used was Saburi. For comparison pearl millet was also grown in another plot as per general recommended package of practices. Weed and pest managements were the same in both the plots.

In Integrated (*Charsutri)* method, number of pearl millet plants were 11 per sq.m., while 14 per sq.m. in recommended practice method. The number of tillers and breadth of earheads were higher in Integrated (*Charsutri)* method, which ultimately increased pearl millet grain yields (Anonymous 2006 b ; Savant, 2007).

Similarly, in *kharif* 2006, 7 trials were conducted on farmers' fields at Jeur, Parewadi and Pargaon in Ahmednagar tahsil and district on Integrated (*Charsutri)* pearl millet farming. Each trial was conducted on 0.40 ha land. They carried out sowing in paired-rows. The soils were light to medium black. One Urea-DAP briquette was deep-placed in between paired-rows only

5-7 cm (2" to 3") depth at every 30 cm (12") distance one week after sowing. Out of 7 trials, two trials were unproductive due to heavy rains and data for 5 trials were available. Fertilizers were applied at 75 percent of recommended level (25 percent less). However, average pearl millet yields obtained were 10.70 qha^{-1} grain and 48.3 qha^{-1} fodder. While in farmers' practice, average yields were 9.2 qha^{-1} grain and 41.4 qha^{-1} fodder. Ahmednagar Tahsil Agriculture Officer informed that Integrated (*Charsutri)* method recorded 17 percent higher yields of pearl millet than that of Farmers' method (Anonymous, 2006 c ; Savant, 2007).

Use of Urea-DAP briquettes required more time and labour charges. But, use of deep placement of briquettes resulted in saving of fertilizers and avoided pollution in climate and reduced weed problem. Thus, expenses on weeding were reduced and grain and fodder yields of pearl millet increased. If the farmers use organic manures along with briquettes, the yields are likely to be further increased. Superintending Agriculture Officer, Ahmednagar district expressed that there is a need for training of farmers regarding paired-row sowing and deep placement of briquettes in the soil.

Farmers response :

Farmers felt that there is a need for conducting more field trials on different soils, seasons, varieties with using of Urea-DAP briquettes for *Charsutri* pearl millet cultivation.

Chapter 13

Integrated use of Urea-DAP Briquettes using Appropriate Irrigations for Different Crops (Only for Research)

13.1 Integrated (*Charsutri*) transplanted rice agrotechnology using appropriate irrigation of 50-100 mm depth (only for research)

Practice 1. Limited and efficient use of organic manures only in furrows for rice cultivation

a. Apply rice hull black ash (or *Bhatura* 1), organic manure (0.5 to 1 kg per sq.m.) only in line seedlings beds by mixing. Then rice seeds are sown in the same line after seed treatment.

b. Use agroforestry approach i.e. spreading *gliricidia* green leaves manuring, approx. 30 kg per R i.e. 3 t ha^{-1} only in furrows (15 cm).

c. Spread small pieces of rice straw (approx. 20 kg per R i.e. 2 t ha^{-1}) on 25 cm ridges (walking lanes) (Fig. 44) (Savant, 2007).

Practice 2. Controlled transplanting of improved or hybrid rice seedlings

Sow improved or hybrid rice seeds on raised beds, 15 cm on both sides of furrows mixed with rice hull ash. Carry out controlled transplanting of 3 to 4 weeks old rice seedlings (2 to 4 seedlings per hill) and 25 hills per sq. m. (Fig. 44).

Practice 3. Deep placement of Urea-DAP briquettes

After controlled transplanting of rice seedlings, carry out deep placement of one Urea-DAP briquette 5 to 10 cm (2" to 4") deep into the soil for every 4 rice hills on the same day until the mud is soft using improved transplanting line rope. In case the holes remain open at the sites of briquettes placement, close the same by hand.

N.B.1 : If small farmers burry rice straw into the soil (as per Practice 1), the requirement of potash is fulfilled. In case potash level in the soil is low as per soil testing, apply potash (K_2O) fertilizer in rice field only in furrows.

N.B. 2 : In case there is incidence of pests and diseases, use integrated pest-disease-weed management. For this purpose use organic (plant origin) and biotic materials, as far as possible.

Practice 4. Use irrigations 50-100 mm depth as per requirement after every 15 to 20 days interval only in furrows

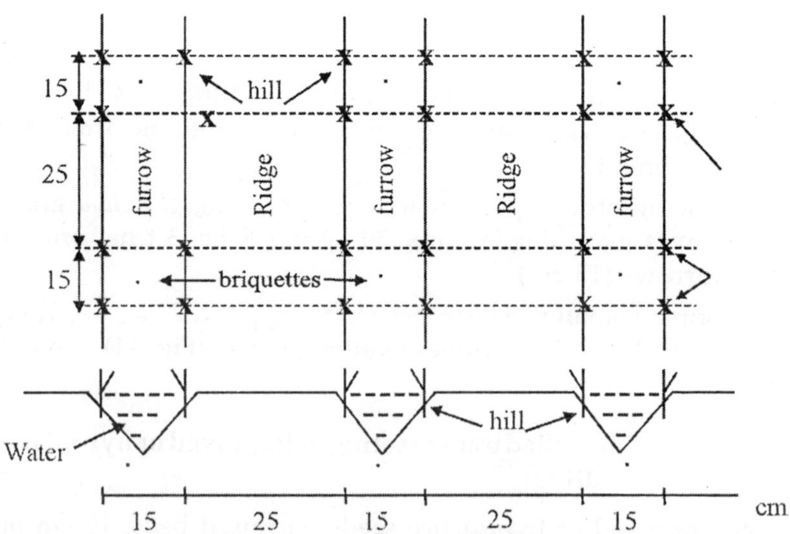

Fig. 44. Controlled transplanting of rice seedlings on both sides of furrows

13.2 Integrated (*Charsutri*) vegetable (organic-inorganic farming) agrotechnology (only for research)

Practice 1. Planting of seedlings/dibbling of seeds in paired-rows by hand

Planting or dibbling is done in south-north direction to get sufficient sunlight (Fig. 45) (Savant, 2007).

Practice 2. Use of limited and efficient organic manures in paired-rows only

Apply 3 to 5 baskets per R (3 to 5 t ha^{-1}) of well decomposed FYM / compost / vermicompost / green manuring of *gliricidia* leaves etc. Use potash fertilizer (20 to 40 kg ha^{-1} K_2O) along with organic manures by mixing.

Practice 3. Limited and efficient use of Urea-DAP briquettes

Place one, two, three or four Urea-DAP briquettes per seedling at 5 to 15 cm away from the seedlings and 5 to 10 cm deep into the soil (20 to 40 percent less fertilizer as compared to general recommended levels, thus saving of expenditure on fertilizers). As per the crop, use briquettes once or in two splits at transplanting and after one to two months of transplanting.

Practice 4. Water management as per requirement

There is saving of water through drip / alternate irrigation.

N.B. : In the case of incidence of pests / diseases, use integrated pest-disease management. If possible, use only plant origin chemicals (insecticides/ fungicides).

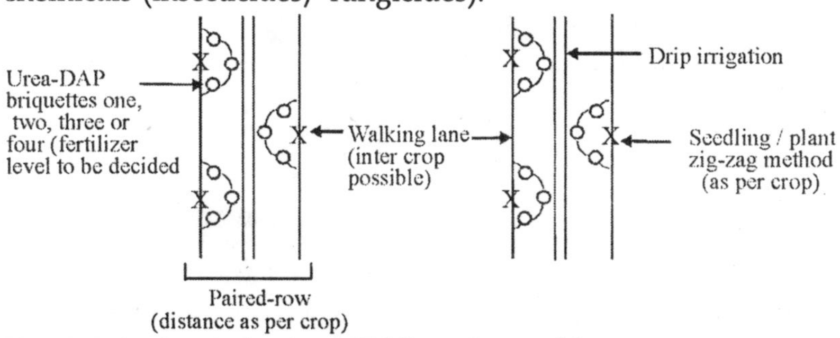

Fig. 45. Paired-row planting / dibbling of vegetables

13.3 Integrated (*charsutri*) sugarcane organic-inorganic farming agrotechnology (only for research)

To accelerate the momentum in agricultural extension in India, there are two aspects of solving farmers' problems : (1) Crop growth in fields is simultaneously influenced by number of environmental factors e.g. local weather, soil conditions and properties, organic-inorganic fertilizers added, pests, diseases and weed problems etc. (Fig. 46). (2) These soil, water and crop practices should not be considered in isolation because of their addition effects (viz. 1+1+1+1~ 4) but if considered in an integrated manner, they have more than total additive effect (viz. 1+1+1+1~ 5 or more) (Fig. 46) (Savant and Sonar, 2007 a).

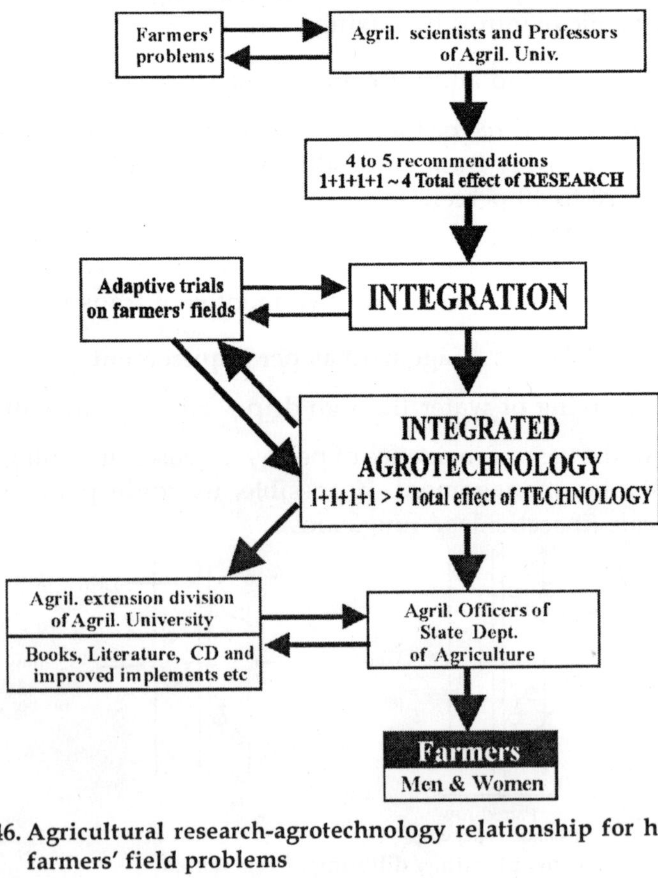

Fig. 46. Agricultural research-agrotechnology relationship for helping farmers' field problems

Integrated sugarcane technology is based on mainly two principles.

i. Integrated plant nutrient system.

ii. Integrated organic- inorganic system.

Integrated sugarcane agrotechnology has 5 practices. Important details are as given below (Savant and Sonar, 2007 a)

Practice 1. Preparation of 1½ to 2 months old one-eye bud sugarcane seedlings in silicon-treated soil (Fig. 47)

a) **Season** : preseasonal sugarcane (Sept.15 to Dec.15)

b) **Cultivars used**: COC-671, CO-8014, CO-86032, CO-219, CO-740 or CO-VSI-9805

c) Selection of planting material:

- Change sugarcane planting material after every three years to preserve the good qualities of the cultivar.
- Germination of the eye-buds should be more than 85%.
- Planting material should be of 8-10 months old.
- Eye-bud should be well developed with thick, strong, fresh looking and free from disease and insect pest attack.
- Do not use ratoon cane, canes with root sprouting and seed formation for planting purpose.
- One-eye bud sets should have 1/3rd upper and 2/3rd lower portion from the bud.

Benefit: (1) This will reduce the cost of healthy and strong planting material by about 66% with increased sugarcane production.

d) Chemical seed treatment :

Dip sugarcane sets in malathion (300 ml per 100 L) and bavistin (100 g per 100 L) for 10-15 min. using lime, $Ca(OH)_2$ (5 kg per 100 L) suspension for better germination of seedlings.

e) **Seedling preparation :**

Two methods of preparation of sugarcane seedlings are as follows :

i. In the same or nearby field, prepare seedlings on seed beds. For planting of one hectare sugarcane, about 7 to 8 R (0.07 to 0.08 ha) area will be required. Prepare field well and 4 to 5 x 1.5 to 2 meter size with 20 to 25 cm raised seed beds. In every bed, 5-6 baskets of compost or FYM is added and mixed with soil. For control of termites and white grubs 10% carbaryl or 2% methyl parathion powder mixed with 100-150 g soil are spread on the beds. For control of stem borer 4% keldon granular insecticide, 50 g per bed mixed. Use keldon after planting.

OR

ii. Prepare seed-treatments with one-eye buds from seedlings in small plastic bags containing soil. (Fig. 47).

Take plastic bags (15 x 15 cm) with 4-5 big holes. Place fine soil + red soil (4:1) + well-decomposed and sieved FYM or compost (2:1) and mix with fine sand. Prepare a mixture of fine soil + red soil (if available) + FYM + fine sand. Add one trolly soil with ½ bag of urea and one bag of single superphosphate and mix well. Apply water every 2-3 days.

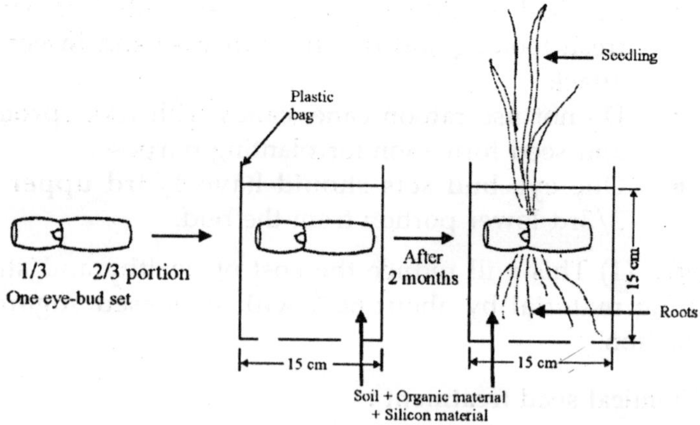

Fig. 47. Preparation of one-eye bud seedlings of sugar in silicon-treated soil

Practice 2. Using south-north direction controlled planting of sugarcane healthy seedlings (75 cm distance paired-rows and 150 cm distance walking lane) in zig-zag manner

a) Planting:

Plant sugarcane in a paired-row (75 cm x 75 cm) with a walking lane of 150 cm. In a paired-row planting, approximately 24000 ha^{-1} of single eye-bud seedlings (sets) are required.

b) Interculturing

First earthing-up: About 2 months after planting, first earthing-up operation is done with the help of bullocks / labourers.

Second (main) earthing-up: Second earthing-up is done 4 to 4.5 months after planting. The ridges are converted into furrows and furrows into ridges in the second earthing-up (Fig. 48).

Benefit: (1) Second earthing-up helps to reduce tillers, remove old unwanted roots, control weeds, reduce lodging of canes and final dose of fertilizers can be applied.

c) Control of tillering:

Approx. 112500 cane number per hectare are required at harvest. For this purpose maintain about 150000 ha^{-1} cane population before final earthing-up.

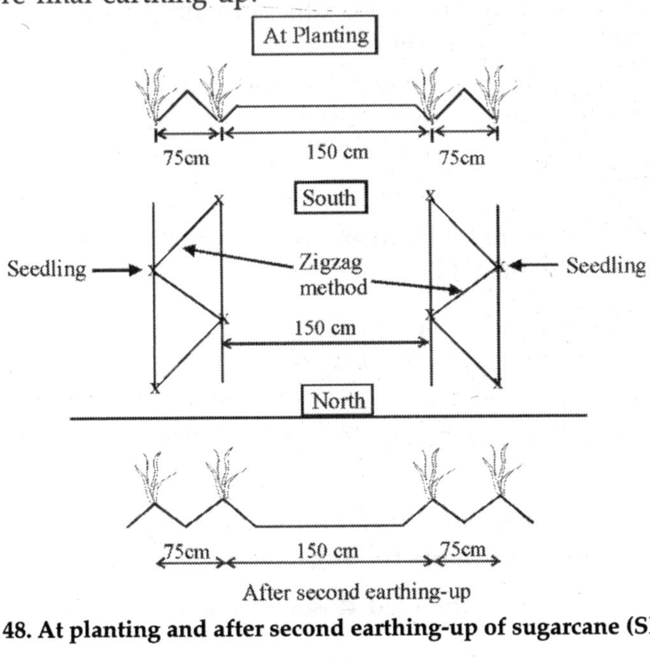

Fig. 48. At planting and after second earthing-up of sugarcane (SISA)

Practice 3. Limited and efficient use of organic and biomanures using silicon containing crop residues recycling for sugarcane

- All organic manures are recommended.
- Prepare land for sugarcane planting, and apply 10 t ha^{-1} of well decomposed and sieved FYM of organic residues.
- At planting of cane, apply remaining 10 t ha^{-1} good quality organic manure (e.g. FYM, compost, vermicompost etc) along with chemical fertilizers in paired-furrows by mixing with soil.
- Apply *gliricidia* green leaf manuring collected from trees near and around the fields as green (leaf) manure @ 5 t ha^{-1} in paired-rows only.

Benefits :

Gliricidia green leaf manuring supplies (1) good quantity green manure every year at minimum cost. (2) wood sticks usedfor household purposes e.g. domestic cooking and (3) protection from rat problem.

- Apply pressmud cake @ 5 t ha^{-1} in paired-rows at second earthing-up operation.
- Use sugar factory ash (e.g. bagasse ash) @ 2 to 4 t ha^{-1} only in paired-rows at second earthing-up operation.

Benefits :

It provides solutions for biotic and abiotic problems of sugarcane (Sonar and Savant, 2007)

Note 1 : For supply of plant available silicon in soil, apply about 500 to 1000 kg ha^{-1} only fly ash pretreated with any organic manures e.g. green *gliricidia* leaves, cattle slurry etc. which needs further research. **Fly ash should be free from any pollution materials.**

Practice 4. Limited and efficient use of N, P, K and micronutrients for sugarcane

Apply 75% of recommended doses of chemical fertilizers for

Western Maharashtra, Marathawada and Vidarbha and other parts of India.

- Use Urea-DAP briquettes (N and P). Place briquettes about 10-15cm deep in the soil with the help of briquette applicator (Figs. 49 and 50).
- Use K (muriate of potash) as per above recommendation along with Ca + S at two growth stages (1) at planting and (2) at second earthing-up.

Benefits :

i. (1) Less weed problem (up to 50%). **(2)** Saving of 25% expenditure on fertilizers. (3) Activity of microorganisms increased.

ii. Integrated disease / insect pest management should be followed. Use plant and animal origin medicines and materials.

iii. Apply micronutrients if the soil is found deficient in micronutrients so as to reduce the cost of cultivation.

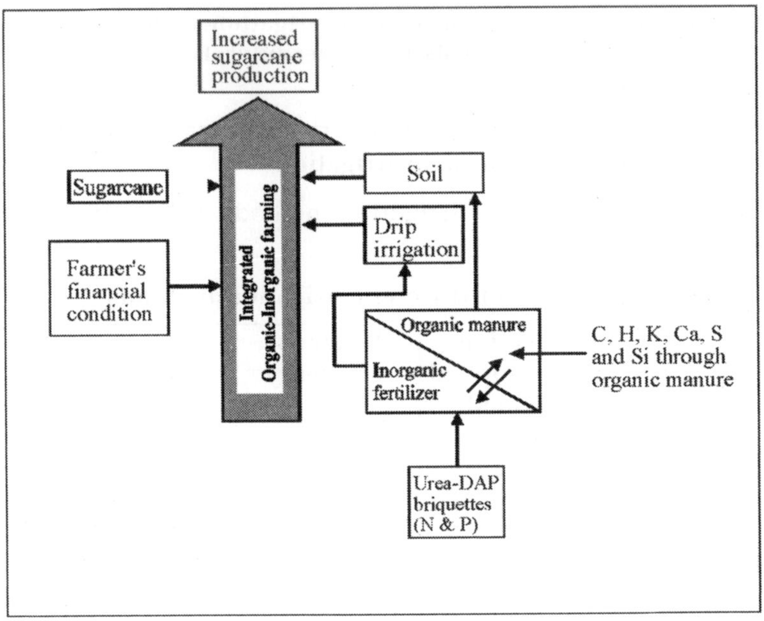

Fig. 49. Drip irrigation using organic manure and inorganic fertilizers

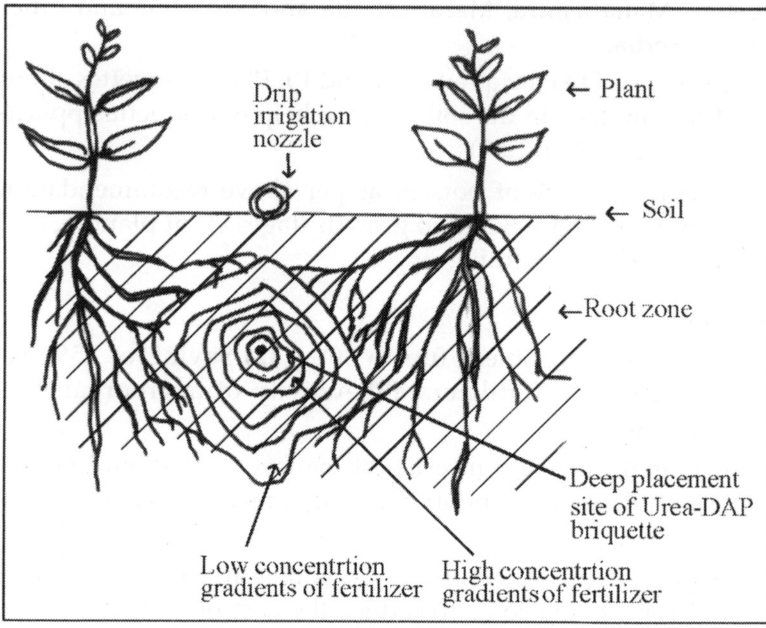

Fig. 50. Schematic diagram of drip irrigation using differential concentration gradients of fertilizer nutrients at placement site of briquette

Practice 5. Efficient methods of irrigation mainly one furrow irrigation or drip irrigation

Efficiencies of different irrigation methods are given in Table 27.

Table 27. Efficiency of different irrigation methods

Irrigation method	Irrigation water applied (ha mm)	Water use efficiency (t ha^{-1})	Cane yield (t ha^{-1})	CCS yield (t ha^{-1})
Sprinkler	1752	0.72	126.5	17.9
Drip	1321	0.97	128.6	18.3
Furrow	2584	0.40	104.4	14.7

- In furrow irrigation and drip irrigation systems using Urea-DAP briquettes (N and P) + drip nozzles , good quality water should result in higher water use efficiency (Nerkar and Patil, 2003 ; Savant, 2007) (Figs. 49 and 50).

EXPECTED BENEFITS OF SISA

1. It is suitable for small farmers with less inputs.
2. It helps to mitigate climatic changes in weather and to prevent diseases and pests of sugarcane.
3. It will help to reduce irrigation and fertilizer costs.

It will increase crop production and will also improve health and productivity of soil.

Chapter 14

Agricultural Implements for Efficient use of Urea-DAP Briquettes in Soil

14.1 Paired-row seed drill

Integrated (*Charsutri*) rice transplanting agrotechnology (integrated organic-inorganic farming agrotechnology) has been popularised among the farmers. Controlled transplanting of seedlings can be performed. In between the two lines of seedlings, Urea-DAP briquettes can be placed efficiently minimising the pollution, increasing the fertilizer use efficiency and increasing rice yields. This agrotechnology has been spread on a large scale among the farmers with the help of Department of Agriculture, Maharashtra Government. Maharashtra Government is providing 25% subsidy for briquette manufacturing machine (Kranti Briquetter) and making the briquettes easily available in the market. At present the briquettes technology for other crops is not available, even then farmers are using briquettes for other crops. Considering the need, Dr. Savant has carried out research on efficient use of briquettes for drilled paddy, wheat, sorghum and pearl millet crops. Deep placement of Urea-DAP briquettes into the root zone of soil eliminates pollution on the farmers' fields. Therefore, Dr. Savant has developed paired-row sowing implement (seed drill) (Photo 29) (Savant, 2007).

14.1.1 Information about improved paired-row seed drill (Photo 29)

i) One plastic funnel

ii) Two plastic tubes connecting the funnel

iii) One handle
iv) Main iron frame
v) Four furrow openers cum coulters
vi) Two tubes carrying seeds from plastic tubes
vii) Iron pipe for pulling of machine

Photo 29. Paired-row seed drill (15 cm spacing with a walking lane of 30 cm)

14.1.2 How to use improved paired-row seed drill?

The farmer who has experience of crop sowing, should take seeds (with seed treatment) in one bag. Seeds are sown as usual by improved seed drill in south-north direction. If the field is small, the seed drill could be used by one worker for dibbling of seeds in paired-rows. If the field is large, controlled paired-row sowing can be done with the help of a pair of bullocks. By using this seed drill, paired-rows at a distance of 15 cm and a walking lane of 30 cm are formed. In the furrows drilled paddy, wheat and sorghum are germinated. Considering the soil slope, as usual channels are prepared for irrigation.

14.1.3 Benefits of using the improved seed drill

i) **Saving of seeds** : 10 to 20 percent less seeds are required.

ii) Helps in crop growth : Considering the soil slope sowing is done in south-north direction with the help of seed drill. This will help in getting sufficient sunlight and indirectly causing healthy crop growth. Because of walking lanes there is a possibility of less incidence of pests and diseases. This reduces expenditure on use of insecticides / fungicides, thus saving on the cost of production with increase in crop yields.

iii) Efficient use of briquettes is possible

Due tó deep placement of (5-7 cm) of Urea-DAP briquettes in soil in paired-rows at 30 cm (12") distance, there are no losses of nutrients from the briquettes. The nutrients in the briquettes become available to the plants for one to two months which is useful for the growth of plants. It is not necessary to apply fertilizers to the crops in 2-3 split doses. Ultimately the crop growth is good. Crop remains fresh, green and healthy resulting in increased crop yields. Fertilizer requirement is reduced by about 20 to 40 percent as compared to usual recommended fertilizer dose. Thus, the expenditure on fertilizer is reduced along with increase in crop yields.

iv) Less weed problem : Due to deep placement of Urea-DAP briquettes, weeds growing on the soil surface do not get nutrients from the fertilizers. Hence the weed problem and eventual labour expenses on weeding are reduced.

14.1.4 Training to the farmers :

Prior to using paired-row seed drill, it is necessary to provide training to the farmers. In the beginning it may take more time for the farmers. But with practice and considering the above benefits, use of paired-row seed drill will become easy.

14.2 Implement (applicator) for deep placement of briquettes in soil for transplanted rice (Vijay-1)

14.2.1 Information about the applicator (Vijay-1) (Photo 30)

The applicator is easy to use and light in weight (portable) (Savant, 2001 a). It is made up of PVC pipe to avoid corrosion with water. As per Integrated (*Charsutri*) transplanted rice agrotechnology, the farmer is adopting controlled transplanting and on the same day in every 4 hills square one Urea-DAP briquette is deep placed in the mud at about 10 cm (4") depth with the help of applicator, Vijay-1. For 1 R (0.01 ha) approximately 625 briquettes (each briquette 2.7 $\pm$ 0.2 g in weight), 1.75 kg (175 kg ha^{-1}) are required.

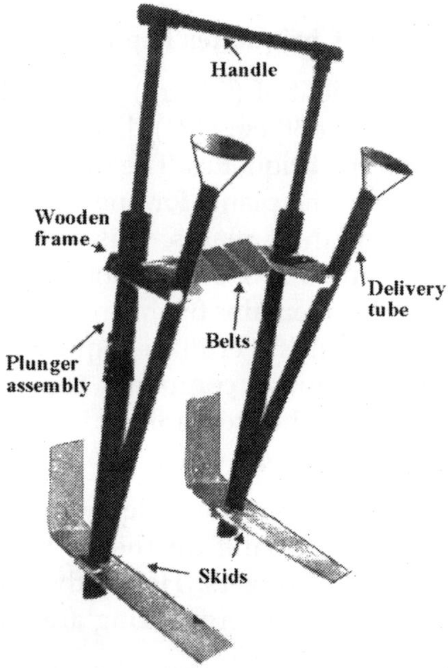

Photo 30. Briquette applicator (Vijay-1) for transplanted rice

14.2.2 Results of demonstration trials on deep placement of Urea-DAP briquettes by applicator for transplanted rice

In all 10 field demonstration trials were conducted successfully on transplanted rice on a scientific basis during 1989-90. Out of these, 5 trials were conducted at International Rice Research Institute (IRRI), Philippines, 3 trials at Agriculture Research Station, Karjat (Maharashtra) and one trial each at Dr. Balasaheb Sawant Kokan Krishi Vidyapeeth, Dapoli (Dist. Ratnagiri) and Central Rice Research Institute, Cuttack (Orisa), respectively (Savant, 2007).

14.2.3 Benefits of using briquette applicator

i) For deep placement of briquettes in the mud, it is not necessary to bow down, thus making the work easy for the farmers.

ii) Expenditure on deep placement of briquettes is reduced. In general, deep placement of briquettes by hand requires 10-12 workers / labourers for one hectare area.

iii) Efficiency of briquettes deep-placed by applicator is the same as that of briquettes deep placed by hand. In fact efficiency of briquettes deep-placed by applicator in good puddled soil has been found to have increased.

iv) Rice grain and straw yields were found to be increased (Table 28) (Savant, 2007 ; Anonymous, 1989 a ; 1989 b ; CRRI, 1989 ; IRRI, 1990).

Table 28. Effect of deep placement of urea briquettes using Vijay-1 applicator on rice yields (1989-90)

Trial	Season	Variety	Avail. N kg ha^{-1}	Rice yield, t ha^{-1}			
				Farmer's method	Recom. method	Briquette Placement by hand	Briquette Placement by applicator
Philippines	1) 1989 summer	IR-64	63	5.81 B	-	7.33 A	7.67 A
	2) 1989 summer	IR-72	63	4.91 C	6.77 B	8.07 A	7.32 A
	3) 1989 *kharif*	IR-72	63	3.59 D	4.12 C	4.44 BC	4.83 A
	4) 1989 *kharif*	IR-72	63	3.30 D	4.31 C	4.81 AB	5.27 A
	5) 1990 summer	IR-72	63	4.23 C	5.53 D	6.23 B	6.60 A
Karjat	6) 1989 *kharif*	Panvel-2	25	2.52 B	2.78 B	3.49 B	3.35 A

	5) 198 9 *kharif*	Panvel-2	40	3.09 C	3.64 B	4.22 A	4.04 A
	8) 1989 *kharif*	Karjat-184	35	2.71 C	3.33 B	4.30 A	3.96 A
Dapoli	6) 198 9 *kharif*	Palghar -1	45	1.67 C	1.79 C	3.98 A	3.27 B
Cuttack	7) 198 9 *kharif*	Gayatri	28	4.63 C	5.03 B	5.33 A	5.26 A

NB : i) In each trial same alphabet next to rice means yields are statistically the same.

ii) In the above trials, rice cultivation and soil conditions after puddling in Philippines were better than that of Indian conditions.

14.2.4 Method of using Vijay-1 implement (briquette applicator)

i) The farmer should complete controlled transplanting of rice seedlings on approximately 2 to 4 R (0.02 to 0.04 ha) area.

ii) One labourer (worker) should discontinue transplanting and start deep placement of Urea-DAP briquettes with the help of Vijay-1 applicator, while other workers should continue transplanting work (Fig. 51) .

iii) The first worker should tie a plastic bag around the waist and take about 1kg of briquettes into the bag.

iv) See that the upward and downward movement of handle pipe of Vijay-1 applicator is easy.

v) Place Vijay-1 applicator in one corner of two paired-rows (in two hills squares) of controlled transplanted rice field. At the centre of the two hills square there should be two wooden frames for deep placement of briquettes. The worker should stand behind the applicator (Fig. 52 A) (Savant and Stangel, 1990 ; Savant *et al.*, 1991 b).

vi) The worker pulls the handle pipe of applicator upward with right hand and at the same time inserts two briquettes in two tubes with left hand (Fig. 52 A).

vii) The worker pushes the handle (plunger) downward for inserting the briquettes into the soil at about 10 cm (4") depth (Fig. 52 B) (since the transplanting of rice was done on the same day, the mud is soft and at the site of briquette placement, the openings are closed with mud).

viii) The worker should pull the handle upward (Fig. 52 C).

ix) After that the worker moves backward by about 40 cm along with Vijay-1 applicator (Fig. 52 D).

This manual briquette placement operation is repeated until briquette placement in two paired-rows is complete. The same worker should keep the applicator in adjoining two paired-rows (Savant, 2001a and 2007). This briquette placement is repeated until the entire field is fertilized.

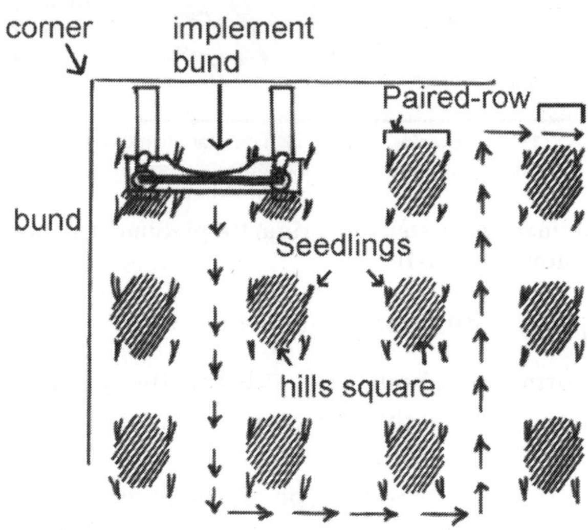

Fig. 51. Method of using Vijay-1 briquette applicator in a small paddy field

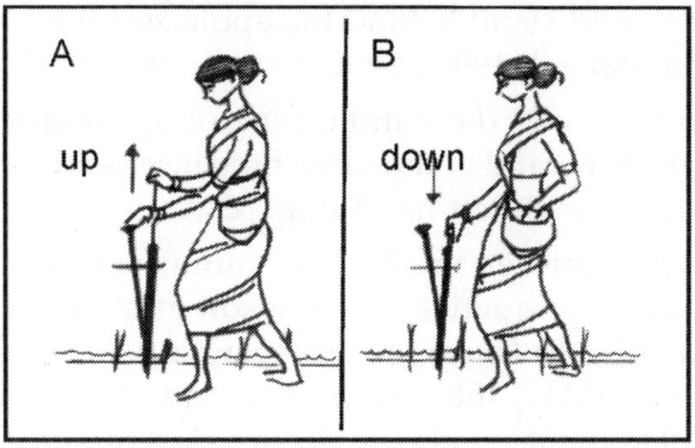

A. A worker places two briquettes, one in each funnel attached to a delivery tube by hand.

B. The worker pushes the handle (plunger) for inserting the briquettes into the soil.

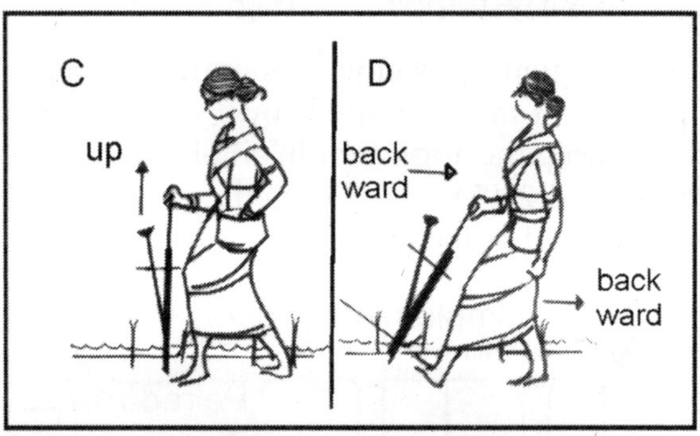

C. The worker pulls the handle upward.

D. The worker moves about 40 cm backward.

Fig. 52. The main four steps of briquette placement using briquette applicator (Vijay-1)

14.3 Briquette applicator implement (Vijay-2)

14.3.1 Information about Vijay-2 briquette applicator

- It is light in weight and easy to operate (Photo 31) (Savant, 2007).
- It is a plunger (handle) type of implement.
- It is prepared by using polyvinyl chloride (PVC) pipes.
- It is hand operated implement.

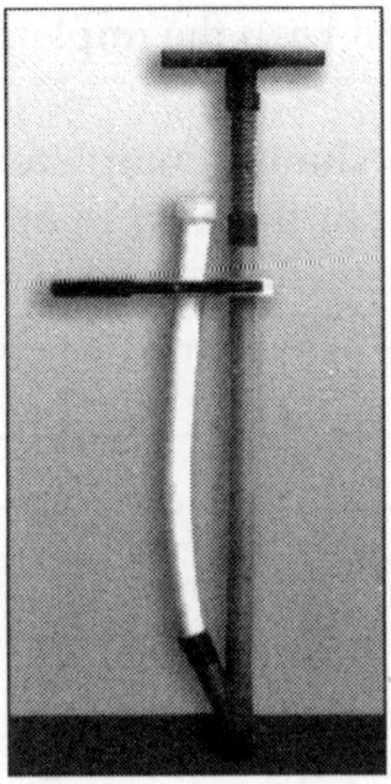

Photo 31. Briquette applicator (Vijay-2)

14.3.2 How to use briquette applicator (Vijay-2)

- After seed germination (10-15 days after sowing), paired-rows can be seen properly. At that time, the applicator could be very well used for deep placement of briquettes for integrated (*Charsutri*) wheat, sorghum, drilled paddy and pearl millet cultivation.
- One worker can tie a plastic bag taking about one kg of briquettes in it.
- The implement is kept anywhere in the paired-rows and the worker should stand behind the applicator.
- See that upward and downward movement of handle bar of Vijay-2 applicator is easy. The worker should take some briquettes in hand and put one briquette in funnel. Then by the other hand push down the handle (plunger) 2-3 times so that the briquette is deep-placed approximately 4-7cm (2"to 3") deep. Due to iron spring, the plunger comes back to its original position. The hole formed at the site of briquette placement could be closed by leg, if necessary.
- Similarly, the same worker should place the briquettes deep into the soil in the front 2 or 3 paired-rows (Figs. 53 and 54).

- One worker should keep the implement about 30 cm (12″) forward.
- Then the worker should complete the operation of briquettes placement in the field in all the paired- rows. After briquette placement, irrigate second time.

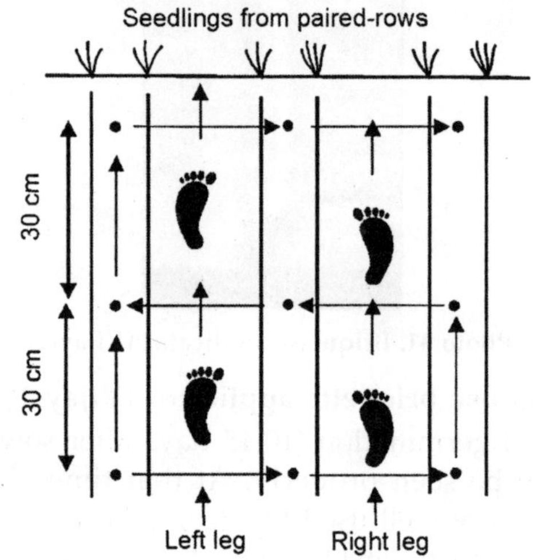

Fig. 53. Method of deep placement of three briquettes in three paired rows by briquette applicator (Vijay-2)

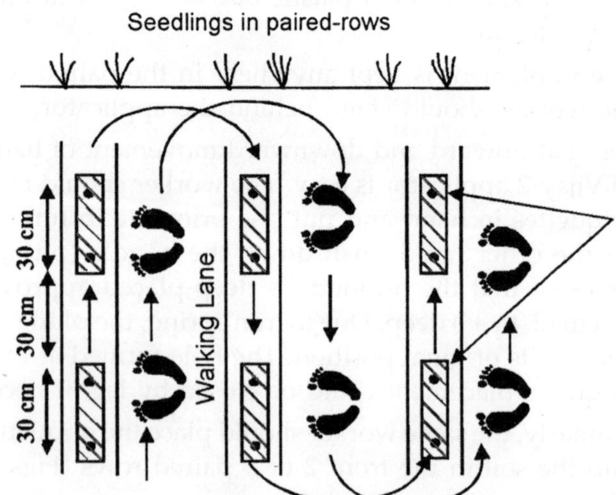

Fig. 54. Method of deep placement of two briquettes in paired-rows by double briquette applicator (Vijay -2)

14.3.3 Precautions to be taken while using applicator

- Use implement only when the soil comes to field capacity condition. If the soil contains more moisture and soil particles come in the plunger, then do not use the implement.
- After using the implement for 20-30 times, clean the plunger and see that deep placement of briquettes is normal.

14.3.4 New implements to be developed

- Briquette applicator - Vijay-3 is being developed and the research is still going on (Photo 32) (Savant, 2007).

Photo 32. Briquette applicator - Vijay-3 being developed for the crops

Chapter 15

Integrated use of Urea-DAP Briquettes with Drip Irrigation System

Organic farming alone is not enough for increasing crop yields. An integrated organic-inorganic farming is gaining increasing importance in India. Extension workers should recommend integrated organic-inorganic farming and popularise it among the farmers. In the case of integrated organic-inorganic farming, efficient use of water is very important in irrigated crops. Stored water from natural rains being a scarce commodity should be used very carefully. This is the need of the hour. For this purpose all efforts should be made to utilize each drop of water efficiently. Agricultural scientists are doing their level best, but they need to look beyond their subjects. Soil, water and crop, their yields and qualities need to be considered in an integrated holistic approach. Considering the method of cultivation of each crop, integrated organic-inorganic farming agrotechnology should be developed. For this purpose considering the soil properties and financial condition of the farmer, drip irrigation system should be established to utilize water efficiently. In drip irrigation management, the farmers need not use water soluble fertilizers but should use Urea-DAP plus briquettes containing N, P, K and other micronutrients. A brief information is given below as to why and how to use briquettes efficiently.

15.1 Benefits of using soluble fertilizers through drip irrigation

There are following benefits of using water soluble fertilizers through drip irrigation by the farmers.

- As per the need of the crop, balanced nutrients are applied in the root zone of the plant.
- There is saving of water soluble fertilizers up to 25 percent than the usual recommended fertilizer levels.
- Fertilizer nutrients are made available to the crops at critical growth stages. Thus, there is good crop growth with yields being increased by about 15 to 25 percent.
- There are more losses of nutrients and atmospheric pollution in the usage of regular fertilizers. However, use of water soluble fertilizers through drip irrigation minimises nutrient losses and pollution.
- Water soluble fertilizers when applied through drip irrigation results in saving of expenditure on labour to some extent. There are several advantages of applying water soluble fertilizers through drip irrigation. Unfortunately very few farmers follow management of water soluble fertilizers through drip irrigation. There is a subsidy from the Maharastra State Government for the use of drip irrigation system, but still majority of farmers are found confused, owing to technical and financial difficulties in use of drip irrigation method. These difficulties must be overcome. In addition to agricultural research, farmers require agricultural technology which is easy, simple and less expensive but still efficient. Only then will the drip irrigation method be popular and accepted by the farmers. There will be economy in use of water with increase in crop yields.

15.2 Technology of using Urea-DAP plus briquettes with drip irrigation

During the year 1992-93, Urea-DAP briquettes were first developed for transplanted rice. Dr. Savant later on developed an idea of using Urea-DAP briquettes with drip irrigation for other crops while he was working with International Fertilizer

Development Centre, U.S.A. In 1994, Dr. Savant prepared briquettes in the laboratory, using Urea-DAP-Potash fertilizer mixture. These briquettes were used with drip irrigation method / underground irrigation method for vegetables viz., tomato, chillie, cabbage and cauliflower in feeler trials. It was found that yields of these crops were increased due to efficient use of nutrients with drip irrigation. Today, many farmers in Maharashtra purchase and use Urea-DAP briquettes through drip irrigation for fruit crops, vegetables, cotton, sugarcane etc. although there is no recommendation. Now, there is a need to carry out research on use of Urea-DAP briquettes with drip irrigation for other crops and to develop technologies for their use.

15.3 The availability of briquettes in Maharashtra

For the spread of integrated (*Charsutri)* transplanted rice farming agrotechnology among the farmers, Maharashtra Government is giving subsidy for purchase of Kranti Briquetter machine for preparing briquettes. These machines are in use all over Maharashtra. At present an average 2.7 gram weight Urea-DAP briquettes are available to the farmers for transplanted rice. For using Urea-DAP-Potash and other micronutrients, 'Urea-DAP plus' briquettes could be prepared and made available in the market. Further research is needed with the help of Maharashtra Government.

15.4 Use of Urea-DAP briquettes with drip irrigation

Urea-DAP plus briquettes can be used with drip irrigation for crops as per the principles given in Figs. 49 and50. Soil remains moist at the site of water dripping from the system. The briquettes are placed deep at this site at about 7 to 10 cm by hand or by implement / applicator and the openings are closed. Thus, the nutrients in the briquettes get dissolved in soil water and remain there, until utilised by the plants. All the nutrients in the briquettes cannot be taken up by the plant roots at one time. This process of utilization of soluble nutrients in water is continued for one to three months (or even more) depending on soil properties and its moisture content.

There is a need for an in depth study on soluble nutrients present in briquettes, their supply to the crops, growth of plant roots etc. While establishing the drip irrigation system there is a need to develop an agrotechnology for each crop regarding the choice of briquettes, time and the method of use.

15.5 The advantages of using Urea-DAP plus briquettes by the farmers

i) Soluble fertilizers have to be imported from other countries. Farmers having irrigation facilities can use soluble fertilizers only for cash crops because of their high cost. The NPK plus briquettes are indigenous and can be prepared from normal fertilizers available in market. The prices of briquettes can be kept reasonable and affordable for the farmers.

ii) The balanced use of Urea-DAP plus briquettes (NPK and other micronutrients) will be easy and cheaper for the farmers as compared to soluble fertilizers.

iii) The components of soluble fertilizer use (drip irrigation system), such as tubes, sieves, pressure measuring devices (machines) etc. are not required while using briquettes. Thus, use of briquettes is less expensive.

iv) If the soluble fertilizers are used, the residue which remains behind choke up / blocks the tubes, per drippers. The farmers will have to clean the tubes which is a trouble some work. If briquettes are used, there is no such problem.

v) The cost and time involved in cleaning of choked tubes in use of soluble fertilizers can be saved by using briquettes instead.

15.6 Need for development of agrotechnology for use of Urea-DAP plus briquettes through drip irrigation

There is a need to first develop a technology within a period of at least 5 years for use of Urea-DAP plus briquettes for crops as a substitute for soluble fertilizers.

i) There is a need to study which cash crops are suitable for the farmers, since drip irrigation system involves a lot of expenditure.

ii) We should plan for the drip irrigation system considering soil type, slope and distance between two lines. The drip irrigation method is better suited for crops having more distance between the lines or paired-row crops (e.g. sugarcane, grapes, fruits, vegetables, flower plants etc.)

iii) The scientists should then organize demonstration trials on farmers' fields on 2-3 crops in 2-3 types of soils for at least two years using Urea-DAP plus briquettes through drip irrigation method. There should be different treatments and replications in these trials. Carry out chemical and statistical analysis for each crop. The agrotechnology can then be developed on the basis of the results of these trials. After developing the agrotechnology for use of Urea-DAP plus briquettes through drip irrigation, popularise it among the farmers.

15.7 Summary

The newer agrotechnology developed by Agricultural Universities / Non Government Ogranisations / Research Institutes and Private Agencies for use of Urea-DAP briquettes through drip irrigation, being simple and inexpensive, many farmers are likely to use this agrotechnology on a large scale, thus bringing more area under crop planting and irrigation.

Chapter **16**

Management of Weeds by using Urea-DAP Briquettes

16.1 Introduction

Integrated (*Charsutri*) transplanted rice farming agrotechnology has been developed recently using integrated nutrient management system. This agrotechnology takes into account balanced nutrient supply and decrease in weed problem in rice crop. For this purpose the farmer uses 3rd and 4th practices out of the 4 management practices of the agrotechnology. Without using any poisonous chemicals, weed problem in the field is reduced with increase in rice yields and less expenditure. The farmer has to adopt first controlled transplanting of rice seedlings, using self prepared *bamboo* guide or improved transplanting line for transplanting of seedlings at two hills and in two lines at a time. Thus, rice seedlings are transplanted at 25 hills per square meter (see Fig. 18). On the day of transplanting in each four hills square, one (average 2.7 gram weight) Urea-DAP briquette is deep-placed. The method of briquette application is easy and very efficient. Due to this method (deep placement of briquettes in soil) of fertilizer application weed management work becomes easy and less expensive.

16.2 Deep placement of briquettes do not promote weed growth

Weed growth is not promoted due to deep placement of Urea-DAP briquettes at about 7 to 10 cm soil depth. Therefore, there is less weed problem (see Fig. 36) (Daftardar *et al.*, 1997a). Broadcasted or incorporated fertilizers remain in floodwater and on surface soil, stimulating weed growth during early crop growth after transplanting. Rice plants have to compete with weeds for nutrients, water and sunlight (see Fig 37). This causes adverse effect on rice plant growth and also yields are reduced. In the case of other crops also (wheat, sorghum and pearl millet etc.), deep placement of Urea-DAP briquettes reduces weed problem.

16.3 Nutrients (N and P) in the briquettes are not wasted as they are not utilized by weeds

Nitrogen and phosphorus in Urea-DAP briquettes are not available to weeds due to deep placement of briquettes in soil. On the other hand, N and P nutrients are slowly utilized by the rice plants as per the need due to deep placement of briquettes in the root zone of soil. Thus, the fertilizers are efficiently utilized for the growth of rice crop (Savant, 2000 ; 2001 a).

16.4 Expenditure on management of weeds is reduced

Weed growth is reduced due to deep placement of Urea-DAP briquettes. This reduces the expenditure on weed management. Many farmers have expressed that one weeding is enough in rice fields as a result of use of briquettes.

16.5 Summary

Deep placement of Urea-DAP briquette for rice reduces weed problem without following any weed management or weedicides. This helps in increasing rice yields. Similarly, use of Urea-DAP briquettes for drilled paddy, wheat, sorghum and pearl millet crops reduces weed problem.

Chapter 17

The Future of Urea-DAP Briquettes (Guidelines)

Food grain production is decreasing in India due to low soil productivity and declining area under agriculture. Thus, food grains production will have to be increased in future by using integrated organic-inorganic farming with Urea-DAP briquettes for increasing soil productivity. Day by day, food problem is becoming critical in India. In future, increase in yields with quality of food will have to be produced from less and less land. In fact this is the need of the hour. For this purpose, integrated use of organic manure with Urea-DAP briquettes needs to be increased.

Suggestions for preparation of briquettes (future guidelines)

i) While using Integrated (*Charsutri)* organic-inorganic farming agrotechnology, there is a need to carry out research on efficient use of briquettes. There is also a need for organising training programmes and field demonstration trials on different crops for agricultural extension workers, fertilizer distributors and men and women farmers.

ii) In case there is deficiency of micronutrients in soil, include these nutrients while preparing the briquettes and develop the technology.

iii) Research will have to be carried out on the use of briquettes for sugarcane, grapes, pomegranate, banana, sapota, orange, sweet orange etc. and vegetables. This will help the farmers to reduce the use of chemical fertilizers.

iv) Do not use urea, DAP and other fertilizers in the field. Instead, use briquettes prepared out of urea polymers and polyphosphates. By using these briquettes it is possible to provide nitrogen and phosphorus in available form for 3 to 6 months for fruit and forest trees.

v) There is a need for conducting research on integrated use of water and Urea-DAP briquettes for different crops and soils. Then develop the agrotechnology which is simple and easy. Thus, the farmer will save water and fertilizers and more area could be brought under drip irrigation.

Summary :

Agricultural Universities, Research Institutes, and State Department of Agriculture and Water Resources Department should take a lead in saving water and fertilizers by conducting research and developing agrotechnologies for integrated use of organic manures, N P briquettes with drip irrigation for different crops.

Chapter 18

Summary

In the year 1930, Japanese agricultural scientists for the first time established that nitrogen fertilizer could be deep-placed in soil for use of rice crop. After that International Fertilizer Development Centre (IFDC) in USA prepared urea supergranules (USG) for use in the fields and conducted demonstration trials on the farmers' fields. In the normal method of fertilizer application for transplanted rice, there are significant losses of nitrogen and phosphorus nutrients. This also causes pollution in soil, water and air. The crop yields are reduced even after applying fertilizers, leading to unprofitable rice farming. Considering all these points, Dr. N.K.Savant developed Urea-DAP briquettes for the economic and efficient supply of N and P for transplanted rice. He developed integrated (*Charsutri*) transplanted rice farming (integrated organic-inorganic farming) agrotechnology for the use of small farmers. This agrotechnology is easy and simple, scientifically efficient, less expensive and pollution free (minimises pollution in soil, water and air). For this purpose he bought technology from IFDC in USA for preparing briquettes. He then prepared average 2.7 gram weight Urea-DAP briquettes (using 60% urea + 40% DAP). Rice yields were increased by using Urea-DAP briquettes. Thousands of field demonstration trials conducted on farmers' fields in Maharashtra and Gujarat States showed that rice yields increased with profitable rice farming. Dr. Savant named the briquetting machine as Kranti Briquetter and the Urea-DAP briquettes as *Vikram*-2 briquettes. Today, Agriculture Department, Maharashtra State is advocating use of integrated (*Charsutri)* transplanted rice farming among the rice farmers. Besides, Agriculture Department of State Government provides 25% subsidy for purchase of Kranti Briquetter machine. Recently,

Dr. Savant also developed integrated (*Charsutri*) organic-inorganic farming agrotechnology for wheat, sorghum and pearl millet crops. This will certainly help in increasing food grain production in the country. For understanding the use of Urea-DAP briquettes for cereal crops, it is necessary to have information on integrated organic-inorganic farming agrotechnology. Similarly, Dr. Savant has developed paired-row seed-drill for easy sowing of the crops by *Charsutri* method (Photo 29). He also developed Vijay-1 and Vijay-2 briquette applicators (Photos 30 and 31). Vijay-3 applicator (Photo 32) is being developed for deep placement of briquettes in the soil. By using Urea-DAP briquettes with drip irrigation system, the farmers can save both water and fertilizer. For this purpose, there is a need to carry out research by Agricultural Universities / Research Institutes for developing easy, simple and less expensive integrated agrotechnology for different crops.

There are several advantages of using Urea-DAP briquettes for the crops, including reduction in weed problem (weed growth) in rice and other crops. Similarly, expenditure on weed management is reduced, as experienced by the farmers who are using Urea-DAP briquettes. Due to adoption of integrated organic-inorganic farming, use of Urea-DAP briquettes is likely to increase in future.

References

Agro One (2008 a) *Development of Rice Grower Farmers at Kadave Village by New Technology*. Tahsil Agriculture Officer, Velhe, Dist. Pune. Daily News Paper, Pune, Feb. 12 (in Marathi).

Agro One (2008 b) *Charsutri Method of Farming is Profitable to Wheat Growers at Malsiras, Dist. Solapur*. Tahsil Agriculture Officer, Malsiras, Solapur District. Daily News Paper, Pune, March 4 (in Marathi).

Agro One (2008 c) *Charsutri Wheat : A Successful Model of Warulwadi, Dist. Pune*. Mandal Agriculture Officer, Narayangaon, Pune District. Daily News Paper, Pune , March 28 (in Marathi).

Agro One (2008 d) *Charsutri Wheat Organic-Inorganic Farming – Propaganda in Ahmednagar District, Rabi 2007-2008* . Joint Director of Agriculture , Pune Region. Daily News Paper, Pune , Oct. 31 (in Marathi).

Agro One (2009) *Co-operative Rice Production Programme Through Improved Technology* . Tahsil Agriculture Officer, Mulshi, Dist. Pune. Daily News Paper, Pune, May, 8 (in Marathi).

Anonymous (1989 a) *Report of Urea Briquette Placement by Applicator (Vijay-1) on Rice Yield, 1989*. Department of Soil Science and Agricultural Chemistry, Dr. Balasaheb Sawant Konkan Krishi Vidyapeeth, Dapoli , Dist. Ratnagiri (unpublished).

Anonymous (1989 b) *Report of Urea Briquette Placement by Applicator (Vijay-1) on Rice Yield, 1989*. Regional Agricultural Research Station, Karjat, Dist. Raigad (unpublished).

Anonymous (1998) *Report of Savant's Integrated Rice Agrotechnology Trials, 1997 and 1998*. Department of Soil Science and Agricultural Chemistry, Dr. Balasaheb Sawant Konkan Krishi Vidyapeeth, Dapoli, Dist. Ratnagiri (unpublished).

Anonymous (1999) *Report of Savant's Integrated Rice Agrotechnology Trials, 1997-1999*. Khar Land Research Station, Panvel , Dist. Raigad (unpublished).

Anonymous (2001 a) *Report of Savant's Integrated Rice Agrotechnology Trials on Farmers' Fields, 1996-2001*. Janaseva Foundation

(Silvassa), Pune (unpublished).

Anonymous (2001 b) *Report of Savant's Integrated Rice Agrotechnology Trials on Farmers' Fields, Kharif 2000.* Superintending Agriculture Officer, Pune District (unpublished).

Anonymous (2001 c) *Report of Savant's Integrated Rice Agrotechnology Trials on Farmers' Fields, Kharif 2000.* Superintending Agriculture Officer , Ahmednagar District (unpublished).

Anonymous (2001 d) *Report of Savant's Integrated Rice Agrotechnology Trials on Farmers' Fields, 1999-2001.* Tahsil Agriculture Officer, Junnar, Pune District (unpublished).

Anonymous (2001 e) *Report of Savant's Integrated Rice Agrotechnology Trials on Farmers' Fields, Kharif 2000.* Superintending Agriculture Officer, Kolhapur District (unpublished).

Anonymous (2001 f) *Report of Savant's Integrated Rice Agrotechnology Trials on Farmers' Fields, Kharif 2000.* Superintending Agriculture Officer, Satara District (unpublished).

Anonymous (2001 g) *Report of Savant's Integrated Rice Agrotechnology Trials on Farmers' Fields, 1999- 2001.* Superintending Agriculture Officer, Nasik District (unpublished).

Anonymous (2001 h) *Report of Savant's Integrated Rice Agrotechnology Trials on Farmers' Fields, Kharif 2001.* Sub-Divisional Agriculture Officer, Dapoli, Dist. Ratnagiri (unpublished).

Anonymous (2001 i) *Report of Savant's Integrated Rice Agrotechnology Trials on Farmers' Fields, Kharif 2001.* Agriculture Development Officer, Zilla Parishad, Kolhapur District (unpublished).

Anonymous (2001 j) *Report of Savant's Integrated Rice Agrotechnology Trials on Farmers' Fields, Kharif 2001.* Block Development Officer, Guhagar, Zilla Parishad, Ratnagiri District (unpublished).

Anonymous (2001 k) *Report of Savant's Integrated Rice Agrotechnology Trials, Kharif 2001.* Regional Agricultural Research Station, Karjat, Dist. Raigad (unpublished).

Anonymous (2001 l) *Report of Savant's Integrated Rice Agrotechnology Trials, Kharif 2001.* Agricultural Research Station, Igatpuri, Dist. Nasik (unpublished).

Anonymous (2001 m) *Report of Savant's Integrated Rice Agrotechnology Trials on Farmers' Fields in Nasik District, Kharif 2000 and 2001.*

Agricultural Research Station, Igatpuri, Dist. Nasik (unpublished).

Anonymous (2001 n) *Report of Savant's Integrated Rice Agrotechnology Trials on Farmers' Fields in Pune District, 2001.* Extension Department, College of Agriculture, Pune (unpublished).

Anonymous (2002) *Report of Savant's Integrated Rice Agrotechnology Trials 1999 to 2002.* Agricultural Research Station, Phondaghat, Dist. Sindhudurg (unpublished).

Anonymous (2003) *Report of Integrated Agrotechnology Trials on upland Rice on Farmers' Fields, 2003, at Radhanagari, Dist. Kolhapur.* Agriculture Development Officer, Zilla Parishad, Kolhapur District (unpublished).

Anonymous (2005 a) *Report of Integrated Agrotechnology Trials on Wheat on Farmers' Fields, 2002-03 to 2004-05 in Aurangabad District.* Superintending Agriculture Officer, Aurangabad Disrtict (unpublished).

Anonymous (2005 b) *Report of Integrated Agrotechnology Trials on Wheat on Farmers' Fields, 2004-05.* Superintending Agriculture Officer, Pune District (unpublished).

Anonymous (2005 c) *Report of Savant's Integrated (Charsutri) Agrotechnology Trials on Rabi, Sorghum on Farmers' Fields, 2004-05, in Pune District.* Superintending Agriculture Officer, Pune District (unpublished).

Anonymous (2005 d) *Report of Savant's Integrated Agrotechnology Trials on Rabi Sorghum on Farmers' Fields, 2004-05, in Barshi Tahsil.* Agricultural Research Station, Barshi, Dist. Solapur (unpublished).

Anonymous (2006 a) *Report of Savant's Integrated Agrotechnology Trials on Rabi Sorghum on Farmers' Fields, 2005-06.* Janaseva Foundation (Silvassa), Pune (unpublished).

Anonymous (2006 b) *Report of Savant's Integrated Agrotechnology Trials on Pearl Millet at Dhule, Rabi 2005-06.* Pearl Millet Scheme, College of Agriculture, Dhule (unpublished).

Anonymous (2006 c) *Report of Savant's Integrated Agrotechnology Trials on Pearl Millet on Farmers' Fields, Kharif 2006, in Ahmednagar District.* Superintending Agriculture Officer, Ahmednagar District (unpublished).

Anonymous (2007 a) *Report of Savant's Integrated Rice Agrotechnology Trials on Farmers' Fields, 2005 to 2007.* Janaseva Foundation (Silvassa), Pune (unpublished).

Anonymous (2007 b) *Report of Integrated Agrotechnology on Wheat on Farmers' Fields, 2006-07, in Surat and Dang Districts.* Janaseva Foundation (Silvassa), Pune (unpublished).

Anonymous (2007 c) *Report of Integrated Agrotechnology Trials on Wheat on Farmers' Fields, 2006-07, in Baramati Tahsil.* Krishi Vigyan Kendra, Baramati , Dist. Pune (unpublished).

Anonymous (2007 d) *Report of Integrated Agrotechnology Trials on Wheat on Farmers' Fields, 2005-06 and 2006-07, in Pune District.* Agricultural School, Manjari, Pune (unpublished).

Anonymous (2007 e) *Report of Savant's Integrated Agrotechnology Trials on Rabi Sorghum on Farmers' Fields, 2005-06 and 2006-07, in Pune District.* College of Agriculture, Pune (unpublished).

Anonymous (2009 a) *Report of Integrated Rice Agrotechnology Trials on Farmers' Fields, 2008.* Janaseva Foundations (Silvassa), Pune (unpublished).

Anonymous (2009 b) *Report of Savant's Integrated Rice Agrotechnology on Farmers' Fields in Ahmednagar District, 2008.* Superintending Agriculture Officer, Ahmednagar District (unpublished).

Anonymous (2009 c) *Report of Savant's Integrated Wheat Agrotechnology on Farmers' Fields, Rabi 2008-09.* Superintending Agriculture Officer, Ahmednagar District (unpublished).

Armstrong W. (1971) Radial oxygen losses from the rice roots as affected by distance from the air respiration and waterlogging. *Physiology of Plant* **25**, 195-197.

Bautista E.U. and Schnier H. F. (1989) Injecting fertilizer solution in wetland paddies. *Agriculture Mechanisation in Asia and Latin America.* **20** (3): 23-27.

Bulbule A.V., Bangal D.B. and Savant N.K. (1995) An efficient N P fertilizer management technique for rainfed transplanted rice in *ghat* zone, Maharashtra. *Fertilizer News* **40** (4): 47-51.

Bulbule A.V. , Talashilkar S.C. and Savant N.K. (1996) Integrated rice straw and urea management for transplanted rice. *Journal of Agriculture Sciences Cambridge* **128**, 291-297.

Cao Z.H., De Datta S. K. and Fillery I.R.P. (1984 a) Effect of placement methods on floodwater properties and recovery of applied N (^{15}N labeled urea) in wetland rice. *Soil Science Society of America Journal* **48**, 196-203.

Cao Z.H., De Datta S. K. and Fillery I.R.P. (1984 b) Nitrogen- 15 balance and residual effects of urea-N in wetland rice fields as affected by deep placement techniques. *Soil Science Society of America Journal* **48**, 203-208.

Chen R.Y., Sun X., Li A., Pan Z., Chen O., and Wei M. (1983) Rational use of nitrogen fertilizer for high yielding rice in Suzhau district. *Acta Pedologica Sinica* **20**, 375-386.

Craswell E.T. and Vlek P. L.G. (1979) Greenhouse evaluation of nitrogen fertilizers for rice. *Soil Science Society of America Journal* **43**, 1184-1188.

Craswell E. T. , De Datta S.K., Obcemea W.N. and Hartantyo M. (1981) Time and mode of nitrogen fertilizer application to tropical wetland rice. *Fertilizer Research* **2**, 247-259.

CRRI (1989) *Report of Urea Briquette Placement by Applicator (Vijay-1) on Rice Yield, Kharif 1989.* Central Rice Research Institute, Cuttack (unpublished).

Daftardar S. Y. (2002) Environment friendly Savant's Integrated Rice Agrotechnology. *Agro India,* August, pp. 18-19.

Daftardar S.Y. and Savant N.K. (1995) Evaluation of environmentally friendly fertilizer management for rainfed lowland rice on tribal farmers' fields in India. *Abstracts of the International Rice Research Conference,* Feb. 13-17. International Rice Research Institute, Los Banos, Laguna, Philippines.

Daftardar S.Y., Wagale S.M. and Savant N.K. (1997 a) Agronomic performance of urea briquettes containing diammonium phosphate in rainfed transplanted rice on farmers' fields. *Journal of Agricultural Sciences, Cambridge* **128**, 291-297.

Daftardar S.Y., Wagale S.M. and Savant N.K. (1997 b) Management of urea briquettes containing diammonium phosphate increases rice yields of small farmers. *International Rice Research Notes* **22** (1): 39-40.

Daftardar S.Y., Wagale S.M. and Savant N.K. (1997 c) Improved N and P fertilizer management increases the rice yields of small resource-poor farmers. *International Rice Commission Newsletter*, FAO, Rome, **46**, 71-72.

De Datta S. K. (1981) *Principles and Practices of Rice Production*. Wiley, New York.

De Datta S. K. (1994) Sustainable rice production : challenges and opportunities. *International Rice Commission Newsletter* **39**, 209-219.

De Datta S. K., Magnaye C. P. and Moomaw J. C. (1968) Efficiency of fertilizer nitrogen (^{15}N labeled) for flooded rice. *Transactions of the 9th International Congress of Soil Science*, Adelaide, Australia, **IV**, pp. 67-76.

De Datta S. K., Sabdaga F. A., Obcemea W. N. and Yoshida T. (1974) Increasing efficiency of fertilizer nitrogen in flooded tropical rice. In : *Proceedings of FAI-FAO Seminar on Optimizing Agricultural Production Under Limited Availability of Fertilizer*, at New Delhi, India, pp. 265-288.

Dhane S. S., Khadse R.R., Patil V.H., Joshi R. G. and Savant N. K. (1991) Agroforestry approach of *gliricidia* green manuring integrated with use of urea supergranules for transplanted rice. *Oryza* **28**, 333-338.

Eriksen A. B. and Kjeldby M. (1987) A comparative study of urea hydrolysis rate and ammonia volatilization from urea and calcium nitrate. *Fertilizer Research* **11**, 9-24

FAO (1984) Maximizing fertilizer use efficiency. *Fertilizer and Plant Nutrition Bulletin No.* **6**, Food and Agriculture Organisation, Rome, Italy.

Holte E., Ryan J. and Ostlyngen T. W. (1982) Pan granulation process for urea supergranules. *Paper Presented at the International Workshop on Nitrogen Management*, April 26-May 4, Fuzhou, China.

IAEA (1970) Rice fertilization. *Technical Report Series No.* **108**. International Atomic Energy Agency, Vienna, Austria.

IFDC (1989) *Small-Scale Briquetter for Producing USG Demonstrated*. International Fertilizer Development Centre, Muscle Shoals, Alabama, USA, Report **12** (4), 1-4.

IFDC (1990) *Annual Report for 1989*. International Fertilizer Development Centre, Muscle Shoals, Alabama, USA.

IRRI (1977) *Research Highlights for 1976*. International Rice Research Institute, Los Banos, Laguna, Philippines, pp. 108.

IRRI (1978 a) *Annual Report for 1977*. International Rice Research Institute, Los Banos, Laguna, Philippines.

IRRI (1978 b) *Soils and Rice*. International Rice Research Institute, Los Banos, Laguna, Philippines.

IRRI (1990) *Report of Urea Briquette Placement by Applicator (Vijay-1)* of *Rice, Summer 1989 and 1990*. International Rice Research Institute, Los Banos, Laguna Philippines (unpublished).

Katyal J. C., Bijay Singh and Craswell E.T. (1984). Fate and efficiency of urea supergranules applied to high percolating rice growing soil. In : *Nitrogen in Soils, Crops and Fertilizers*. Indian Society of Soil Science, New Delhi, Bulletin **13**, 229-237.

Khan A. U. (1984) Deep placement fertilizer application project. *Semi-Annual Progress Report No.* **1**, International Rice Research Institute , Los Banos, Laguna, Philippines.

Kimuta M., Wada H. and Takai Y. (1979) The studies on rhizosphere of paddy rice. vi. The effect of anaerobics on microbes. *Soil Science and Plant Nutrition* **25**, 145-151.

Kumar V., Shrotriya G. C. and Kaore S. V. (Eds.) (1989) *Urea Supergranules for Increasing Nitrogen Use Efficiency*. Indian Farmers Fertilizers Co-Operative Ltd., New Delhi, India, pp. 1-143.

Kumbhar C. T., Nevase A. G. and Savant N. K. (1995) Rice hull ash applied to soil reduces leaf blast incidence. *International Rice Research Notes* **20**, 23-24.

Kumbhar C. T. and Savant N. K. (2001 a) Recycling of rice plant silicon and potassium for blast management in rice. In : *Silicon in Agriculture*, (Datnoff L.E., Snyder G.H. and Korndorfer G.H., Eds.), Elsevier Science, Amsterdam, (*Proceedings of Seminar, Sept. 26-30, 1999*, at Fort Lauderdale, USA), pp.382.

Kumbhar C. T. and Savant N. K. (2001 b) Recycling of rice plant silicon and potassium for leaf scald management in rice. In : *Silicon in Agriculture*, (Datnoff L.E., Snyder G.H. and Korndorfer G.H., Eds.), Elsevier Science, Amsterdam, (*Proceedings of Seminar, Sept. 26-30, 1999*, at Fort Lauderdale, USA), pp.382-383.

Kunitaka M. (1962) *Studies on the Deep Placement of Urea Subsequent to Transplanting* . Republic Niigata Agriculture Experiment Station,

13 (in Japanese).

Kyuma K. (1989) Rice soils of Asia : distribution and management. In: *Classification and Management of Rice Growing Soils*. Food and Fertilizer Technology Centre, Taipei, Taiwan, Republic of China, pp. 15-17.

Li C. K. and Chen R. Y. (1980) Ammonium bicarbonate used as a nitrogen fertilizer in China. *Fertilizer Research* **1**, 125-136.

Lindau C. W., Reddy K. R., Lu W., Khind C. S., Pardue J. H. and Patrick W. H. Jr. 1989. Effect of redox potential on urea hydrolysis and nitrification in soil suspension. *Soil Science* **148**, 184-190.

Lupin M. S. (1984) *Technical Bulletin, T* **26** . International Fertilizer Development Centre, Muscle Shoals, Alabama, USA.

Mengel K., Schon H. G., Keerthisinghe G. and De Datta S. K. (1986) Ammonium dynamics of puddled soils in relation to growth and yield of lowland rice. *Fertilizer Research* **9**, 117-130.

Misra C., Mohanty B. C., Das B. S. and Savant N. K. (1995) Relationship between some selected soil properties and yield of transplanted rice fertilized with urea briquettes. *Oryza* **32**, 178-183.

Mitsui S. (1977) Recognition of the importance of denitrification and its impact on various improved and mechanised applications of nitrogen to rice plant. In : *Soil Environment and Fertility Management in Intensive Agriculture . Proceedings of Seminar*, Society of Science of Soils and Manure, Tokyo, Japan, pp. 259-269.

Modgal S. C., Singh Y. and Gupta P. C. (1995) Nutrient management in rice-wheat cropping system. *Fertilizer News* **40** (5) 49-54.

Nerkar Y. S. and Patil A. S. (Eds.) (2003) *Sugarcane Cultivation Technology*. Vasantdada Sugar Institute, Pune, pp. 1-89 (in Marathi).

Panda D., Sen H. S. and Patnaik S. (1988) Spatial and temporal distribution of nitrogen in a puddled rice soil following application of urea-based fertilizer by different methods. *Biology and Fertility of Soils* **6**, 89-92.

Panda N. and Patnaik S. 1989. Fate and efficiency of urea supergranules in different rice culture systems in deltai alluvial soil of eastern India. *Soil Fertility and Fertilizer Use* **3**, 58-66, Indian Farmers FertilizersCo-perative Ltd., New Delhi, India.

Pingali P. L. (1993) Technological and policy option for sustaining rice productivity growth in Asia. *Food Fertilizer Technical Centre Extension Bulletin No.* **377**, Taipei, Taiwan.

Ponnaiyya J.H.S. and Varadarajan G. (1979) *Gravitor Applicator : An Equipment for Placement of Fertilizers and Insecticides in Rice Field.* Director of Agriculture, Chepauk, Madras, India, pp. 1-10.

Ponnamperuma F.N. (1972) The chemistry of submerged soils. *Advances in Agronomy* **24**, 29-96.

Powar S. L. and Deshpande V. N. (2000) Effect of integrated agrotechnology on Sahyadri hybrid rice in medium black soil in high rainfall area. *Paper Presented at State Level Seminar on Soil Technology for Sustainable Agriculture,* at Mahatma Phule Krishi Vidypeeth, Rahuri, Dist. Ahmednagar, Jan. 18-19.

Pulford I. D. and Tabatabai M. A. (1988) Effect of waterlogging on enzyme activities in soil. *Soil Biology and Biochemistry*, 215-219.

Rao M. V. (1983) Studies with supergranules for increasing the fertilizer efficiency in rice. In : *Proceedings of Seminar-cum-Workshop on Urea Supergranules* . Indian Farmers Fertilizers Co-Operative, Ltd., New Delhi, and University of Agricultural Sciences, Bangalore, May 27-28, 1982, pp.24-36.

Reddy K. R. and Patrick W. H. Jr. (1986) Denitrification losses in flooded rice soils. *Fertilizer Research* **9**, 99-116.

Roger P. A., Kulasooriya S. A., Tirol A. C. and Craswell E. T. (1980) Deep placement : A method of nitrogen fertilizer application compatible with algal nitrogen fixation in wetland rice soils. *Plant and Soil* **57**, 137-140.

Russell R. S. (1977) *Plant Root Systems, Their Functions and Interaction with Soil* . McGraw Hill, New York.

Savant N. K. (1979) Fate of deep-placed urea in wetland rice soil. *Summary Report August 1977-April 1979 (Mimeograph),* Agron. Dept., IRRI, Los Banos, Laguna, Philippines.

Savant N. K. (1995) Increased rice production with less fertilizer. *Information Centre for Low-External Input and Sustainable Agriculture (ILEIA) Newsletter* **11** (1): 24-25.

Savant N. K. (2000) *Savant's Charsutri Rice Farming for Small Farmers.* Continental Publication, Pune, pp. 1-126 (in Marathi).

Savant N. K. (2001a) *A Guide to Savant's Integrated Rice Agrotechnology.*

Continental Publication, Pune pp. 1-117.

Savant N. K. (2001 b) Urea-DAP briquettes preparing machine. *Shetkari Magazine* **2** (2): 43-44 (in Marathi).

Savant N. K. (Ed.) (2002) *Proceedings of the State Level Seminar on Savant's Integrated Rice Agrotechnology*, at Pune, March 7, pp. 1-54.

Savant N. K. (2005) Savant's integrated rice agrotechnology for farmers of Maharashtra : a review. *Journal of Maharashtra Agricultural Universities* **30** (2), 127-130.

Savant N. K. (2006) *Integrated (Charsutri) Agrotechnology of Cereal Crops Farming for Small Farmers* . Krishi Granth Bhandar Publication, Pune, pp. 1-150 (in Marathi).

Savant N. K. (2007) *Vikram-2 Briquettes for Crops (Urea-DAP briquettes)*. Agro One Publication, Pune, pp. 1-73 (in Marathi).

Savant N. K. and De Datta S. K. (1979) Nitrogen release pattern from deep placement sites of urea in a wetland rice soil. *Soil Science Society of America Journal* **43**, 131-134.

Savant N. K. and De Datta S. K. (1980) Movement and distribution of ammonium-N following deep placement of urea in a wetland rice soil. *Soil Science Society of America Journal* **44**, 559-565.

Savant N. K. and De Datta S. K. (1982) Nitrogen transformations in wetland rice soils. *Advances in Agronomy* **35**, 241-302.

Savant N. K., De Datta S. K. and Craswell E. T. (1982) Distribution patterns of ammonium nitrogen and ^{15}N uptake by rice after deep placement of urea supergranules in wetland soil. *Soil Science Society of America Journal* **46**, 567-573.

Savant N. K., Craswell E. T. and Diamond R. B. (1983 a). Use of urea supergranules for wetland rice : A review. *Fertilizer News* **28** (8): 27-35.

Savant N. K., James A. F. and McChallan G. H. (1983 b). Effect of soil submergence on urea hydrolysis. *Soil Science* **140**, 81-88.

Savant N. K. and Stangel P. J. (1985) Suggestions on efficient use of urea supergranules for wetland rice. *Proceedings of the Workshop on Urea Placement Technology*, pp. 133-143 (Special Publication of International Fertilizer Development Centre , Muscle Shoals, AL, USA and Agency of Agricultural Research and Development (AARD), Bogor, Indonesia).

Savant N. K. and Chien S. H. (1990) Greenhouse evaluation of urea supergranules containing diammonium phosphate for transplanted rice. *International Rice Research Newsletter* **15**(3): 23.

Savant N. K. and Stangel P. J. (1990). Deep placement of urea supergranules in transplanted rice : principles and practices. *Fertilizer Research* **25**, 1-83.

Savant N. K., Dhane S. S., Talashilkar S. C. and Patil N. S. (1991 a) Rainfed rice yields response to urea supergranules deep-placed during line transplanting. *Fertilizer News* **36** (3): 19-25.

Savant N. K., Ongkingco P. S., Zarate I. V., Torrizo F. M. and Stangel P. J. (1991 b) Urea briquette applicator for transplanted rice. *Fertilizer Research* **28**, 323-331.

Savant N. K., Ongkingco, Garcia F. D., Dhane S. S., Khadse R. R., Chavan S.S. and Rao K. S. (1992) Agronomic performance of urea briquette applicator in transplanted rice. *Fertilizer Research* **32**, 139-142.

Savant N. K., Talashilkar S. C., Dalvi A. S. and Dhane S. S. (1993) Transplanted rice yields are influenced by deep-placed urea briquettes fortified with diammonium phosphate on lateritic soils of Kokan. *Abstracts of International Seminar on Managing Red and Lateritic Soils for Sustainable Agriculture*, at Bangalore, Sept. 24-28.

Savant N. K., Carmona G. and Austin E. R. (1994) Effect of plant density on methane emmission from transplanted rice. *International Rice Research Notes* **19** (4): 24-25.

Savant N. K. and Sawant A. S. (1995) Nutrient composition of rice seedlings as influenced by application of rice hull ash to seedbed. *Oryza* **33**, 74-76.

Savant N. K. and Stangel P. J. (1995) Recent developments in urea briquettes use for transplanted rice. *Fertilizer News* **40** (8): 27-33.

Savant N. K. and Stangel P. J. (1996) Recent developments in use of urea briquettes for transplanted rice. *Fertilizer News* **40** (8): 29-33.

Savant N. K., Datnoff L. E. and Snyder G. H. (1997 a) Depletion of plant available silicon in soil : a possible cause of declining rice yields. *Communication in Soil Science and Plant Analysis* **20**, 195-201.

Savant N. K., Snyder G. H. and Datnoff L. E. (1997 b) Silicon management and sustainable rice production. *Advances in*

Agronomy **58**, 151-199.

Savant N. K., Menon R. G. and Friesen D. K. (1997 c) Transplanting geometry improves timing of uptake of deep point-placed P by rice hills. *International Rice Research Notes* **22**(1): 36.

Savant N. K. and Stangel P. J. (1998) Urea-briquettes containing diammonium phosphate : A potential new N P fertilizer for transplanted rice. *Nutrient Cycling in Agroecosystems* **51**, 85-94.

Savant N. K. and Daftardar S. Y. (2000) *Savant's Integrated Rice Agrotechnology for Small Farmers : A review*. Booklet No. **4** , Janaseva Foundation (Silvassa), Pune, pp 24.

Savant N. K., Daftardar S. Y. and Ingale B. V. (2004) Savant's integrated rice agrotechnology for more hybrid rice production with less pollution. *Paper Presented at Zonal Conference on Hybrid Rice : Problems and Prospects*, Nov. 29-30, at Dr. Balasaheb Savant Konkan Krishi Vidyapeeth, Dapoli, Maharashtra.

Savant N. K. and Sonar K. R. (2007 a) Savant's integrated sugarcane agrotechnology : a conceptual outline. *Proceedings of the National Seminar on Fertigation in Sugarcane, Dec. 19-20,* at Vasantdada Sugar Institute, Pune, pp. 60-69.

Savant N. K. and Sonar K. R. (2007 b) *Silicon : A Tonic for Rice and Sugarcane* . Krishi Granth Bhandar Publication, Pune, pp 146 (in Marathi).

Sawant A. S., Patil V. H. and Savant N. K. (1994) Rice hull ash applied to seedbed reduces deadhearts in transplanted rice. *International Rice Research Notes* **19** (4): 21-22.

Schnier H. F., De Datta S. K., Mengel K. and Marqucos I. and Faronilo J. E. (1988) Nitrogen use efficiency, floodwater properties and nitrogen-15 balance in transplanted lowland rice as affected by liquid urea deep placement. *Fertilizer Research* **16**, 241-255.

Sistani K. R., Savant N. K. and Reddy K. C. (1997) Effect of rice hull ash silicon on rice seedling growth. *Journal of Plant Nutrition* **20** (1): 195-201.

Sonar K. R. and Savant N. K. (2007) Integrated nutrient management in relation to silicon for increasing sugarcane production. *Proceedings of the National Seminar on Fertigation in Sugarcane,* Dec. 19-20, Vasantdada Sugar Institute, Pune, pp. 70-85.

Stangel P. J. (1989) Urea supergranules for transplanted rice : IFDC research review. *Soil Fertility and Fertilizer Use* **3**, 1-9, Indian Farmers Fertilizers Co-Operative Ltd., New Delhi, India.

Stangel P. J. and Harris G. T. (1987) Trends in production, trade and use of fertilizers : a global perspective. In : *Efficiency of Nitrogen Fertilizers for Rice* . International Rice Research Institute , Los Banos, Laguna, Philippines, pp. 1-26

Stangel P. J., Savant N. K. and Byrnes B. H. (1993) Potential of modified ureas for rice. In : *New Frontiers in Rice Research* (Muralidharan and Siddiq Eds.). Directorate of Rice Research, Hyderabad Publication, pp. 300-307.

Talashilkar S. C. and Savant N. K. (1991) Response of rice to deep-placed urea supergranules in Ultisols of Maharashtra, India. In : *Rice Production on Acid Soils of the Tropics* (Deturck and Ponnamperuma Eds). Institute of Fundamental Studies, Kandy, Sri Lanka, pp.181-184.

Talashilkar S. C., Sawant A. S., Dharmapurkar V. B. and Savant N. K. (2000) Yield and nutrient uptake by rice as influenced by integrated use of calcium silicate slag and UB-DAP in acid lateritic soil. *Journal of the Indian Society of Soil Science* **48**, 847-849.

Vachhani M. V. (1952) Fertilizer use for stepping up of rice production. *Indian Farming* **2** (1): 28-31.

Van der Sar T. and de Vries H. C. P. (1984) A urea point injector for wetland rice. In : *Fertilizer Application Equipment for Small Farmers.* Food and Agriculture Organisation, Rome, Italy, pp.87-93.

Vlek P. L. G., Stumpe J. M. and Byrnes B. H. (1980). Urease activity and inhibition in flooded soil systems. *Fertilizer Research* **1**, 191-202.

Vlek P. L. G. and Byrnas B. H. (1986) The efficiency and loss of fertilizer N in lowland rice. *Fertilizer Research* **9**, 131-147.

Vyas B. N. and Mistry K. B. (1985) Hydrolysis of prilled urea, sulphur-coated urea and urea supergranules in Oxisol, Vertisol and Inceptisol. *Indian Journal of Agricultural Sciences* **55**, 35-40.

Watanabe I. and Mitsui S. (1979) Denitrification loss of fertilizer nitrogen in paddy soils - its recognition and impact. *IRRI Research Paper Series* **37**, pp. 10.

Wetselaar R. (1985) Deep point placed urea in flooded soil : a mechanistic view. *Proceedings of the Urea Deep Placement Technology Workshop*, Sept. 1984 at Bogor, Indonesia, pp 7-15.

Yadav A.N., Dhane S.S. and Mahale Dilip. (2002) Effect of method, plant density and nutrient management on grain yield of rice under coastal saline condition. *Paper Presented at State Level Seminar on Land Resources Management, Food Security and Permanency*, Dr. Balasaheb Sawant Konkan Krishi Vidyapeeth, Dapoli, Dist. Ratnagiri.

Appendix - 1

Definition : Slow-release fertilizers and slowly available fertilizers.

i) Slow-release fertilizers :

e.g. Sulphur coated urea. Solubility of urea is dependent on less or more layers coated on urea granules. These fertilizers when applied to soil, react with soil moisture within few minutes. Urea fertilizer starts dissolving depending on the plastic coating given. In soil moisture, urea-N is converted into nitrate-nitrogen (NO_3-N) which is available to the plant roots slowly. Crop growth is increased which ultimately increases crop yields. Because of high cost, the slow-release fertilizers are not affordable to the general farmers. At present, these fertilizers are used for lawns in USA, Japan and other rich countries. Similar other slow-release fertilizers are rubber coated urea, polymer coated urea, polyurethane coated urea etc.

ii) Slowly available fertilizers

e.g. urea supergranules, Urea-DAP briquettes. Urea-DAP briquettes is a mixed fertilizer (34: 17:0). Deep placement of briquettes in soil is necessary for their efficient use. After application of briquettes in soil it reacts with soil moisture and gets dissolved completely. That means these briquettes are not slow release fertilizers. Then their higher concentration gradients to lower concentration gradients are formed (Figs. 2 and 3). First, plant roots take-up nutrients from lower concentration gradient solution. Then the nutrients move from higher concentration gradient to lower concentration gradient. The movement of nutrients from higher to lower concentration gradients continues according to the soil moisture content. During the movement of nutrients, each time the plant roots absorb nutrients from low concentration gradient solution. Thus, the process of supply of nutrients is continued for one to two months and the crop roots get nutrient supply up to 90%. The crop gets nutrient supply in required proportion and whenever required. The crop growth is increased which ultimately increases the crop yields.

Index